Dean Chalmers
Series Editor: Julian Gilbey

Cambridge International
AS & A Level Mathematics:

Probability & Statistics 1

Coursebook

CAMBRIDGE
UNIVERSITY PRESS

University Printing House, Cambridge CB2 8BS, United Kingdom

One Liberty Plaza, 20th Floor, New York, NY 10006, USA

477 Williamstown Road, Port Melbourne, VIC 3207, Australia

314–321, 3rd Floor, Plot 3, Splendor Forum, Jasola District Centre, New Delhi – 110025, India

103 Penang Road, #05-06/07, Visioncrest Commercial, Singapore 238467

Cambridge University Press is part of the University of Cambridge.

It furthers the University's mission by disseminating knowledge in the pursuit of education, learning and research at the highest international levels of excellence.

Information on this title:
www.cambridge.org/9781108407304 (Paperback)
www.cambridge.org/9781108462242 (Cambridge Online Mathematics, 2 years)
www.cambridge.org/9781108610827 (Paperback + Cambridge Online Mathematics, 2 years)

© Cambridge University Press 2018

This publication is in copyright. Subject to statutory exception
and to the provisions of relevant collective licensing agreements,
no reproduction of any part may take place without the written
permission of Cambridge University Press.

First published 2018

20 19 18 17 16 15 14 13 12 11 10 9

Printed in Poland by Opolgraf

A catalogue record for this publication is available from the British Library

ISBN 978-1-108-40730-4 Paperback
ISBN 978-1-108-61082-7 Paperback + Cambridge Online Mathematics, 2 years
ISBN 978-1-108-46224-2 Cambridge Online Mathematics, 2 years

Cambridge University Press has no responsibility for the persistence or accuracy of URLs for external or third-party internet websites referred to in this publication, and does not guarantee that any content on such websites is, or will remain, accurate or appropriate. Information regarding prices, travel timetables, and other factual information given in this work is correct at the time of first printing but Cambridge University Press does not guarantee the accuracy of such information thereafter.

® *IGCSE is a registered trademark*

Past exam paper questions throughout are reproduced by permission of Cambridge Assessment International Education. Cambridge Assessment International Education bears no responsibility for the example answers to questions taken from its past question papers which are contained in this publication.

The questions, example answers, marks awarded and/or comments that appear in this book were written by the author(s). In examination, the way marks would be awarded to answers like these may be different.

NOTICE TO TEACHERS IN THE UK
It is illegal to reproduce any part of this work in material form (including photocopying and electronic storage) except under the following circumstances:
(i)   where you are abiding by a licence granted to your school or institution by the Copyright Licensing Agency;
(ii) where no such licence exists, or where you wish to exceed the terms of a licence, and you have gained the written permission of Cambridge University Press;
(iii) where you are allowed to reproduce without permission under the provisions of Chapter 3 of the Copyright, Designs and Patents Act 1988, which covers, for example, the reproduction of short passages within certain types of educational anthology and reproduction for the purposes of setting examination questions.

Contents

Series introduction — vi

How to use this book — viii

Acknowledgements — x

1 Representation of data — 1
1.1 Types of data — 2
1.2 Representation of discrete data: stem-and-leaf diagrams — 3
1.3 Representation of continuous data: histograms — 6
1.4 Representation of continuous data: cumulative frequency graphs — 13
1.5 Comparing different data representations — 20
End-of-chapter review exercise 1 — 24

2 Measures of central tendency — 26
2.1 The mode and the modal class — 28
2.2 The mean — 30
2.3 The median — 42
End-of-chapter review exercise 2 — 51

3 Measures of variation — 54
3.1 The range — 55
3.2 The interquartile range and percentiles — 56
3.3 Variance and standard deviation — 65
End-of-chapter review exercise 3 — 83

Cross-topic review exercise 1 — 86

4 Probability — 90
4.1 Experiments, events and outcomes — 91
4.2 Mutually exclusive events and the addition law — 94
4.3 Independent events and the multiplication law — 100
4.4 Conditional probability — 108
4.5 Dependent events and conditional probability — 112
End-of-chapter review exercise 4 — 118

5 Permutations and combinations — 122
5.1 The factorial function — 124
5.2 Permutations — 125

5.3	Combinations	135
5.4	Problem solving with permutations and combinations	138
	End-of-chapter review exercise 5	143

Cross-topic review exercise 2 146

6 Probability distributions 149

6.1	Discrete random variables	150
6.2	Probability distributions	150
6.3	Expectation and variance of a discrete random variable	156
	End-of-chapter review exercise 6	162

7 The binomial and geometric distributions 165

7.1	The binomial distribution	166
7.2	The geometric distribution	175
	End-of-chapter review exercise 7	185

8 The normal distribution 187

8.1	Continuous random variables	188
8.2	The normal distribution	193
8.3	Modelling with the normal distribution	205
8.4	The normal approximation to the binomial distribution	208
	End-of-chapter review exercise 8	215

Cross-topic review exercise 3 217

Practice exam-style paper 220

The standard normal distribution function 222

Answers 223

Glossary 245

Index 247

Series introduction

Cambridge International AS & A Level Mathematics can be a life-changing course. On the one hand, it is a facilitating subject: there are many university courses that either require an A Level or equivalent qualification in mathematics or prefer applicants who have it. On the other hand, it will help you to learn to think more precisely and logically, while also encouraging creativity. Doing mathematics can be like doing art: just as an artist needs to master her tools (use of the paintbrush, for example) and understand theoretical ideas (perspective, colour wheels and so on), so does a mathematician (using tools such as algebra and calculus, which you will learn about in this course). But this is only the technical side: the joy in art comes through creativity, when the artist uses her tools to express ideas in novel ways. Mathematics is very similar: the tools are needed, but the deep joy in the subject comes through solving problems.

You might wonder what a mathematical 'problem' is. This is a very good question, and many people have offered different answers. You might like to write down your own thoughts on this question, and reflect on how they change as you progress through this course. One possible idea is that a mathematical problem is a mathematical question that you do not immediately know how to answer. (If you do know how to answer it immediately, then we might call it an 'exercise' instead.) Such a problem will take time to answer: you may have to try different approaches, using different tools or ideas, on your own or with others, until you finally discover a way into it. This may take minutes, hours, days or weeks to achieve, and your sense of achievement may well grow with the effort it has taken.

In addition to the mathematical tools that you will learn in this course, the problem-solving skills that you will develop will also help you throughout life, whatever you end up doing. It is very common to be faced with problems, be it in science, engineering, mathematics, accountancy, law or beyond, and having the confidence to systematically work your way through them will be very useful.

This series of Cambridge International AS & A Level Mathematics coursebooks, written for the Cambridge Assessment International Education syllabus for examination from 2020, will support you both to learn the mathematics required for these examinations and to develop your mathematical problem-solving skills. The new examinations may well include more unfamiliar questions than in the past, and having these skills will allow you to approach such questions with curiosity and confidence.

In addition to problem solving, there are two other key concepts that Cambridge Assessment International Education have introduced in this syllabus: namely communication and mathematical modelling. These appear in various forms throughout the coursebooks.

Communication in speech, writing and drawing lies at the heart of what it is to be human, and this is no less true in mathematics. While there is a temptation to think of mathematics as only existing in a dry, written form in textbooks, nothing could be further from the truth: mathematical communication comes in many forms, and discussing mathematical ideas with colleagues is a major part of every mathematician's working life. As you study this course, you will work on many problems. Exploring them or struggling with them together with a classmate will help you both to develop your understanding and thinking, as well as improving your (mathematical) communication skills. And being able to convince someone that your reasoning is correct, initially verbally and then in writing, forms the heart of the mathematical skill of 'proof'.

Series introduction

Mathematical modelling is where mathematics meets the 'real world'. There are many situations where people need to make predictions or to understand what is happening in the world, and mathematics frequently provides tools to assist with this. Mathematicians will look at the real world situation and attempt to capture the key aspects of it in the form of equations, thereby building a model of reality. They will use this model to make predictions, and where possible test these against reality. If necessary, they will then attempt to improve the model in order to make better predictions. Examples include weather prediction and climate change modelling, forensic science (to understand what happened at an accident or crime scene), modelling population change in the human, animal and plant kingdoms, modelling aircraft and ship behaviour, modelling financial markets and many others. In this course, we will be developing tools which are vital for modelling many of these situations.

To support you in your learning, these coursebooks have a variety of new features, for example:

- Explore activities: These activities are designed to offer problems for classroom use. They require thought and deliberation: some introduce a new idea, others will extend your thinking, while others can support consolidation. The activities are often best approached by working in small groups and then sharing your ideas with each other and the class, as they are not generally routine in nature. This is one of the ways in which you can develop problem-solving skills and confidence in handling unfamiliar questions.
- Questions labelled as **P**, **M** or **PS**: These are questions with a particular emphasis on 'Proof', 'Modelling' or 'Problem solving'. They are designed to support you in preparing for the new style of examination. They may or may not be harder than other questions in the exercise.
- The language of the explanatory sections makes much more use of the words 'we', 'us' and 'our' than in previous coursebooks. This language invites and encourages you to be an active participant rather than an observer, simply following instructions ('you do this, then you do that'). It is also the way that professional mathematicians usually write about mathematics. The new examinations may well present you with unfamiliar questions, and if you are used to being active in your mathematics, you will stand a better chance of being able to successfully handle such challenges.

At various points in the books, there are also web links to relevant Underground Mathematics resources, which can be found on the free **undergroundmathematics.org** website. Underground Mathematics has the aim of producing engaging, rich materials for all students of Cambridge International AS & A Level Mathematics and similar qualifications. These high-quality resources have the potential to simultaneously develop your mathematical thinking skills and your fluency in techniques, so we do encourage you to make good use of them.

We wish you every success as you embark on this course.

Julian Gilbey
London, 2018

Past exam paper questions throughout are reproduced by permission of Cambridge Assessment International Education. Cambridge Assessment International Education bears no responsibility for the example answers to questions taken from its past question papers which are contained in this publication.

The questions, example answers, marks awarded and/or comments that appear in this book were written by the author(s). In examination, the way marks would be awarded to answers like these may be different.

How to use this book

Throughout this book you will notice particular features that are designed to help your learning. This section provides a brief overview of these features.

In this chapter you will learn how to:
- display numerical data in stem-and-leaf diagrams, histograms and cumulative frequency graphs
- interpret statistical data presented in various forms
- select an appropriate method for displaying data.

Learning objectives indicate the important concepts within each chapter and help you to navigate through the coursebook.

KEY POINT 1.2

Data in a stem-and-leaf diagram are ordered in rows of equal widths.

Key point boxes contain a summary of the most important methods, facts and formulae.

A **cumulative frequency graph**

Key terms are important terms in the topic that you are learning. They are highlighted in orange bold. The **glossary** contains clear definitions of these key terms.

EXPLORE 3.5

The following table shows three students' marks out of 20 in the same five tests.

	1st	2nd	3rd	4th	5th	
Amber	12	17	11	9	16	x
Buti	11	16	10	8	15	$x-1$
Chen	15	20	14	12	19	$x+3$

Note that Buti's marks are consistently 1 less than Amber's and that Chen's marks are consistently 3 more than Amber's. This is indicated in the last column of the table.

For each student, calculate the variance and standard deviation.

Can you explain your results, and do they apply equally to the range and interquartile range?

Explore boxes contain enrichment activities for extension work. These activities promote group-work and peer-to-peer discussion, and are intended to deepen your understanding of a concept. (Answers to the Explore questions are provided in the Teacher's Resource.)

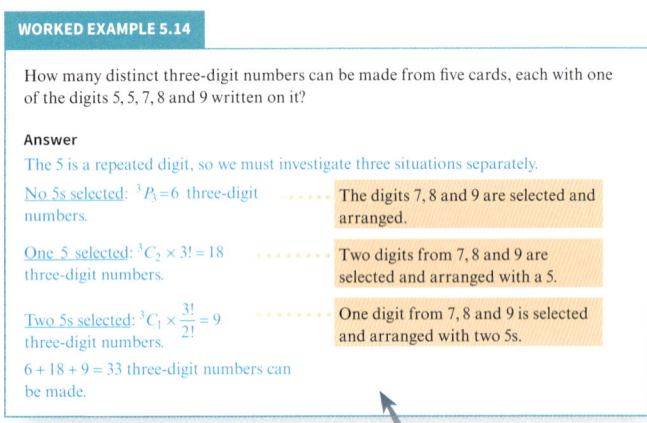

Prerequisite knowledge exercises identify prior learning that you need to have covered before starting the chapter. Try the questions to identify any areas that you need to review before continuing with the chapter.

WORKED EXAMPLE 5.14

How many distinct three-digit numbers can be made from five cards, each with one of the digits 5, 5, 7, 8 and 9 written on it?

Answer

The 5 is a repeated digit, so we must investigate three situations separately.

No 5s selected: $^3P_3 = 6$ three-digit numbers. ····· The digits 7, 8 and 9 are selected and arranged.

One 5 selected: $^3C_2 \times 3! = 18$ three-digit numbers. ····· Two digits from 7, 8 and 9 are selected and arranged with a 5.

Two 5s selected: $^3C_1 \times \dfrac{3!}{2!} = 9$ three-digit numbers. ····· One digit from 7, 8 and 9 is selected and arranged with two 5s.

$6 + 18 + 9 = 33$ three-digit numbers can be made.

Worked examples provide step-by-step approaches to answering questions. The left side shows a fully worked solution, while the right side contains a commentary explaining each step in the working.

 TIP

A variable is denoted by an upper-case letter and its possible values by the same lower-case letter.

Tip boxes contain helpful guidance about calculating or checking your answers.

How to use this book

> **REWIND**
>
> Recall from Chapter 4, Section 4.1 that $P(A) = 1 - P(A')$.

> ⏭ **FAST FORWARD**
>
> Later in this section, we will see how any normal variable can be transformed to the standard normal variable by coding.

Extension material goes beyond the syllabus. It is highlighted by a red line to the left of the text.

Rewind and **Fast forward** boxes direct you to related learning. **Rewind** boxes refer to earlier learning, in case you need to revise a topic. **Fast forward** boxes refer to topics that you will cover at a later stage, in case you would like to extend your study.

> ⓘ **DID YOU KNOW?**
>
> It has long been common practice to write *Q.E.D.* at the point where a mathematical proof or philosophical argument is complete. *Q.E.D.* is an initialism of the Latin phrase *quad erat demonstrandum*, meaning 'which is what had to be shown'.
>
> Latin was used as the language of international communication, scholarship and science until well into the 18th century.
>
> *Q.E.D.* does not stand for *Quite Easily Done*!
>
> A popular modern alternative is to write W^5, an abbreviation of *Which Was What Was Wanted*.

Did you know? boxes contain interesting facts showing how Mathematics relates to the wider world.

> **Checklist of learning and understanding**
> - Commonly used measures of variation are the range, interquartile range and standard deviation.
> - A box-and-whisker diagram shows the smallest and largest values, the lower and upper quartiles and the median of a set of data.

At the end of each chapter there is a **Checklist of learning and understanding** and an **End-of-chapter review exercise**.

The checklist contains a summary of the concepts that were covered in the chapter. You can use this to quickly check that you have covered the main topics.

Cross-topic review exercises appear after several chapters, and cover topics from across the preceding chapters.

Throughout each chapter there are multiple exercises containing practice questions. The questions are coded:

- **PS** These questions focus on problem-solving.
- **P** These questions focus on proofs.
- **M** These questions focus on modelling.
- 🚫🖩 You should not use a calculator for these questions.
- 🖩 You can use a calculator for these questions.
- 📄 These questions are taken from past examination papers.

> **END-OF-CHAPTER REVIEW EXERCISE 8**
>
> 1. A continuous random variable, X, has a normal distribution with mean 8 and standard deviation σ. Given that $P(X > 5) = 0.9772$, find $P(X < 9.5)$. [3]
> 2. The variable Y is normally distributed. Given that $10\sigma = 3\mu$ and $P(Y < 10) = 0.75$, find $P(Y \geq 6)$ [4]
> 3. In Scotland, in November, on average 80% of days are cloudy. Assume that the weather on any one day is independent of the weather on other days.

The **End-of-chapter review** contains exam-style questions covering all topics in the chapter. You can use this to check your understanding of the topics you have covered. The number of marks gives an indication of how long you should be spending on the question. You should spend more time on questions with higher mark allocations; questions with only one or two marks should not need you to spend time doing complicated calculations or writing long explanations.

> **CROSS-TOPIC REVIEW EXERCISE 2**
>
> 1. Each of the eight players in a chess team plays 12 games against opponents from other teams. The tota wins, draws and losses for the whole team are denoted by X, Y and Z, respectively.
>   a. State the value of $X + Y + Z$.

Acknowledgements

The authors and publishers acknowledge the following sources of copyright material and are grateful for the permissions granted. While every effort has been made, it has not always been possible to identify the sources of all the material used, or to trace all copyright holders. If any omissions are brought to our notice, we will be happy to include the appropriate acknowledgements on reprinting.

Past examination questions throughout are reproduced by permission of Cambridge Assessment International Education.

Thanks to the following for permission to reproduce images:
Cover image Mint Images – Danita Delimont/Getty Images
Inside *(in order of appearance)* Barcroft/Contributor/Getty Images, Rick Etkin/Getty Images, Plume Creative/Getty Images, Artography/Shutterstock, Kypros/Getty Images, Anadolu Agency/Contributor/Getty Images, SensorSpot/Getty Images, Andrew Spencer/Contributor/Getty Images, Spencer Platt/Staff/Getty Images, ATTA KENARE/Staff/Getty Images, Pablo Blazquez Dominguez/Stringer/Getty Images, Fred Ramage/Stringer/Getty Images, Garry Gay/Getty Images, De Agostini Picture Library/Getty Images, Frederic Woirgard/Look At Sciences/Science Photo Library, Echo/Getty Images, By Yáng Huī, ca. 1238–1298 [Public domain], via Wikimedia Commons, INDRANIL MUKHERJEE/Staff/Getty Images, Heritage Images/Contributor/Getty Images

Chapter 1
Representation of data

In this chapter you will learn how to:

- display numerical data in stem-and-leaf diagrams, histograms and cumulative frequency graphs
- interpret statistical data presented in various forms
- select an appropriate method for displaying data.

Cambridge International AS & A Level Mathematics: Probability & Statistics 1

PREREQUISITE KNOWLEDGE

Where it comes from	What you should be able to do	Check your skills

IGCSE® / O Level Mathematics	Obtain appropriate upper and lower bounds to solutions of simple problems when given data to a specified accuracy.	1 A rectangular plot measures 20 m by 12 m, both to the nearest metre. Find:
		   a its least possible perimeter
		   b the upper boundary of its area.

IGCSE / O Level Mathematics	Construct and interpret histograms with equal and unequal intervals.	2 A histogram is drawn to represent two classes of data. The column widths are 3 cm and 4 cm, and the column heights are 8 cm and 6 cm, respectively. What do we know about the frequencies of these two classes?

IGCSE / O Level Mathematics	Construct and use cumulative frequency diagrams.	3 The heights of 50 trees are measured: 17 trees are less than 3 m; 44 trees are less than 4 m; and all of the trees are less than 5 m. Determine, by drawing a cumulative frequency diagram, how many trees have heights:
		   a between 3 and 4 m
		   b of 4 m or more.

Why do we collect, display and analyse data?

We can collect data by gathering and counting, taking surveys, giving out questionnaires or by taking measurements. We display and analyse data so that we can describe the things, both physical and social, that we see and experience around us. We can also find answers to questions that might not be immediately obvious, and we can also identify questions for further investigation.

Improving our data-handling skills will allow us to better understand and evaluate the large amounts of statistical information that we meet daily. We find it in the media and from elsewhere: sports news, product advertisements, weather updates, health and environmental reports, service information, political campaigning, stock market reports and forecasts, and so on.

Through activities that involve data handling, we naturally begin to formulate questions. This is a valuable skill that helps us to make informed decisions. We also acquire skills that enable us to recognise some of the inaccurate ways in which data can be represented and analysed, and to develop the ability to evaluate the validity of someone else's research.

1.1 Types of data

There are two types of data: **qualitative** (or **categorical**) **data** are described by words and are non-numerical, such as blood types or colours. **Quantitative data** take numerical values and are either discrete or continuous. As a general rule, **discrete data** are counted and cannot be made more precise, whereas continuous data are measurements that are given to a chosen degree of accuracy.

Discrete data can take only certain values, as shown in the diagram.

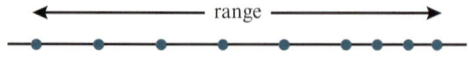

The number of letters in the words of a book is an example of discrete quantitative data. Each word has 1 or 2 or 3 or 4 or… letters. There are no words with $3\frac{1}{3}$ or 4.75 letters.

Discrete quantitative data can take non-integer values. For example, United States coins have dollar values of 0.01, 0.05, 0.10, 0.25, 0.50 and 1.00. In Canada, the United Kingdom and other countries, shoe sizes such as $6\frac{1}{2}$, 7 and $7\frac{1}{2}$ are used.

Continuous data can take any value (possibly within a limited range), as shown in the diagram.

The times taken by the athletes to complete a 100-metre race is an example of continuous quantitative data. We can measure these to the nearest second, tenth of a second or even more accurately if we have the necessary equipment. The range of times is limited to positive real numbers.

> **KEY POINT 1.1**
>
> Discrete data can take only certain values.
> Continuous data can take any value, possibly within a limited range.

1.2 Representation of discrete data: stem-and-leaf diagrams

A **stem-and-leaf diagram** is a type of table best suited to representing small amounts of discrete data. The last digit of each data value appears as a *leaf* attached to all the other digits, which appear in a *stem*. The digits in the stem are **ordered** vertically, and the digits on the leaves are ordered horizontally, with the smallest digit placed nearest to the stem.

Each row in the table forms a **class** of values. The rows should have intervals of equal width to allow for easy visual comparison of sets of data. A **key** with the appropriate unit must be included to explain what the values in the diagram represent.

Stem-and-leaf diagrams are particularly useful because **raw data** can still be seen, and two sets of related data can be shown back-to-back for the purpose of making comparisons.

Consider the raw percentage scores of 15 students in a Physics exam, given in the following list: 58, 55, 58, 61, 72, 79, 97, 67, 61, 77, 92, 64, 69, 62 and 53.

To present the data in a stem-and-leaf diagram, we first group the scores into suitable equal-width classes.

Class widths of 10 are suitable here, as shown below.

```
5 | 8 5 8 3
6 | 1 7 1 4 9 2
7 | 2 9 7
8 |
9 | 7 2
```

Next, we arrange the scores in each row in ascending order from left to right and add a key to produce the stem-and-leaf diagram shown below.

```
5 | 3 5 8 8        Key: 5 | 3
6 | 1 1 2 4 7 9    represents
7 | 2 7 9          a score of 53%
8 |
9 | 2 7
```

> **TIP**
>
> The diagram should have a bar chart-like shape, which is achieved by aligning the leaves in columns. It is advisable to redraw the diagram if any errors are noticed, or to complete it in pencil, so that accuracy can be maintained.

In a back-to-back stem-and-leaf diagram, the leaves to the right of the stem ascend left to right, and the leaves on the left of the stem ascend right to left (as shown in Worked example 1.1).

Cambridge International AS & A Level Mathematics: Probability & Statistics 1

WORKED EXAMPLE 1.1

The number of days on which rain fell in a certain town in each month of 2016 and 2017 are given.

Year 2016					
Jan: 17	Feb: 20	Mar: 13	Apr: 12	May: 10	Jun: 8
Jul: 0	Aug: 1	Sep: 5	Oct: 11	Nov: 16	Dec: 9

Year 2017					
Jan: 9	Feb: 13	Mar: 11	Apr: 8	May: 6	Jun: 3
Jul: 1	Aug: 2	Sep: 2	Oct: 4	Nov: 8	Dec: 7

Display the data in a back-to-back stem-and-leaf diagram and briefly compare the rainfall in 2016 with the rainfall in 2017.

Answer

```
     2016   |   | 2017
 9 8 5 1 0  | 0 | 1 2 2 3 4 6 7 8 9
 7 6 3 2 1 0| 1 | 1 3
          0 | 2 |
```

Key: 5 | 0 | 6 represents 5 days in a month of 2016 and 6 days in a month of 2017

We group the values for the months of each year into classes 0–9, 10–19 and 20–29, and then arrange the values in each class in order with a key, as shown.

It rained on more days in 2016 (122 days) than it did in 2017 (74 days).

No information is given about the amount of rain that fell, so it would be a mistake to say that more rain fell in 2016 than in 2017.

> **TIP**
>
> If rows of leaves are particularly long, repeated values may be used in the stem (but this is not necessary in Worked example 1.1). However, if there were, say, 30 leaves in one of the rows, we might consider grouping the data into narrower classes of 0–4, 5–9, 10–14, 15–19 and 20–24. This would require 0, 0, 1, 1 and 2 in the stem.

> **KEY POINT 1.2**
>
> Data in a stem-and-leaf diagram are ordered in rows of equal widths.

EXERCISE 1A

1. Twenty people leaving a cinema are each asked, "How many times have you attended the cinema in the past year?" Their responses are:

   6, 2, 13, 1, 4, 8, 11, 3, 4, 16, 7, 20, 13, 5, 15, 3, 12, 9, 26 and 10.

   Construct a stem-and-leaf diagram for these data and include a key.

2. A shopkeeper takes 12 bags of coins to the bank. The bags contain the following numbers of coins:

   150, 163, 158, 165, 172, 152, 160, 170, 156, 162, 159 and 175.

   a Represent this information in a stem-and-leaf diagram.

   b Each bag contains coins of the same value, and the shopkeeper has at least one bag containing coins with dollar values of 0.10, 0.25, 0.50 and 1.00 only.

   What is the greatest possible value of all the coins in the 12 bags?

3. This stem-and-leaf diagram shows the number of employees at 20 companies.

   ```
   1 | 0 8 8 8 8 9 9       Key: 1 | 0
   2 | 0 5 6 6 7 7 8 9     represents 10
   3 | 0 1 1 2 9           employees
   ```

   a What is the most common number of employees?

   b How many of the companies have fewer than 25 employees?

c   What percentage of the companies have more than 30 employees?

d   Determine which of the three rows in the stem-and-leaf diagram contains the smallest number of:

   i   companies   ii   employees.

4   Over a 14-day period, data were collected on the number of passengers travelling on two ferries, A and Z. The results are presented to the right.

Ferry A (14)		Ferry Z (14)
8 7 6	2	
7 6 4 0	3	0 5 8
8 6 5 3	4	3 4 5 7 7 7
5 3 3	5	0 2 6 6 9

Key: 3 | 5 | 0 represents 53 passengers on A and 50 passengers on Z

a   How many more passengers travelled on ferry Z than on ferry A?

b   The cost of a trip on ferry A is $12.50 and the cost of a trip on ferry Z is $x. The takings on ferry Z were $3.30 less than the takings on ferry A over this period. Find the value of x.

c   Find the least and greatest possible number of days on which the two ferries could have carried exactly the same number of passengers.

5   The runs scored by two batsmen in 15 cricket matches last season were:

Batsman P: 53, 41, 57, 38, 41, 37, 59, 48, 52, 39, 47, 36, 37, 44, 59.

Batsman Q: 56, 48, 31, 64, 21, 52, 45, 36, 57, 68, 77, 20, 42, 51, 71.

a   Show the data in a diagram that allows easy comparison of the two performances.

b   Giving a reason for your answer, decide which of the batsmen performed:

   i   better   ii   more consistently.

6   The total numbers of eggs laid in the nests of two species of bird were recorded over several breeding seasons.

The numbers of eggs laid in the nests of 10 wrens and 10 dunnocks are:

Wrens: 22, 18, 21, 23, 17, 23, 20, 19, 24, 13.

Dunnocks: 28, 24, 23, 19, 30, 27, 22, 25, 22, 17.

a   Represent the data in a back-to-back stem-and-leaf diagram with rows of width 5.

b   Given that all of these eggs hatched and that the survival rate for dunnock chicks is 92%, estimate the number of dunnock chicks that survived.

c   Find the survival rate for the wren chicks, given that 14 did not survive.

PS 7   This back-to-back stem-and-leaf diagram shows the percentage scores of the 25 students who were the top performers in an examination.

Girls (12)		Boys (13)
4 1	8	2
8 6 6	8	5 9
3 2 1 0	9	0 1 3 3 4 4
8 7 7	9	5 6 6 9

Key: 1 | 8 | 2 represents 81% for a girl and 82% for a boy

The 25 students are arranged in a line in the order of their scores. Describe the student in the middle of the line and find the greatest possible number of boys in the line who are not standing next to a girl.

1.3 Representation of continuous data: histograms

Continuous data are given to a certain degree of accuracy, such as 3 significant figures, 2 decimal places, to the nearest 10 and so on. We usually refer to this as *rounding*.

When values are rounded, gaps appear between classes of values and this can lead to a misunderstanding of continuous data because those gaps do not exist.

Consider heights to the nearest centimetre, given as 146–150, 151–155 and 156–160.

Gaps of 1cm appear between classes because the values are rounded.

Using h for height, the actual classes are $145.5 \leq h < 150.5$, $150.5 \leq h < 155.5$ and $155.5 \leq h < 160.5$ cm.

The classes are shown in the diagram below, with the **lower and upper boundary** values and the **class mid-values** (also called midpoints) indicated.

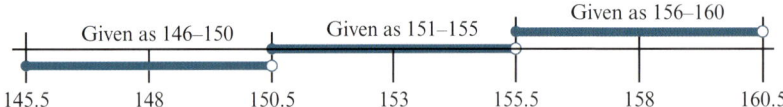

Lower **class boundaries** are 145.5, 150.5 and 155.5 cm.

Upper class boundaries are 150.5, 155.5 and 160.5 cm.

Class widths are $150.5 - 145.5 = 5$, $155.5 - 150.5 = 5$ and $160.5 - 155.5 = 5$.

Class mid-values are $\frac{145.5 + 150.5}{2} = 148$, $\frac{150.5 + 155.5}{2} = 153$ and $\frac{155.5 + 160.5}{2} = 158$.

A **histogram** is best suited to illustrating continuous data but it can also be used to illustrate discrete data. We might have to group the data ourselves or it may be given to us in a **grouped frequency table**, such as those presented in the tables below, which show the ages and the percentage scores of 100 students who took an examination.

Age (A years)	$16 \leq A < 18$	$18 \leq A < 20$	$20 \leq A < 22$
No. students (f)	34	46	20

Score (%)	10–29	30–59	60–79	80–99
No. students (f)	6	21	60	13

> **TIP**
> 'No.' is the abbreviation used for 'Number of' throughout this book.

The first table shows three classes of continuous data; there are no gaps between the classes and the classes have equal-width intervals of 2 years. This means that we can represent the data in a frequency diagram by drawing three equal-width columns with column heights equal to the class frequencies, as shown below.

> **TIP**
> We *concertina* part of an axis to show that a range of values has been omitted.

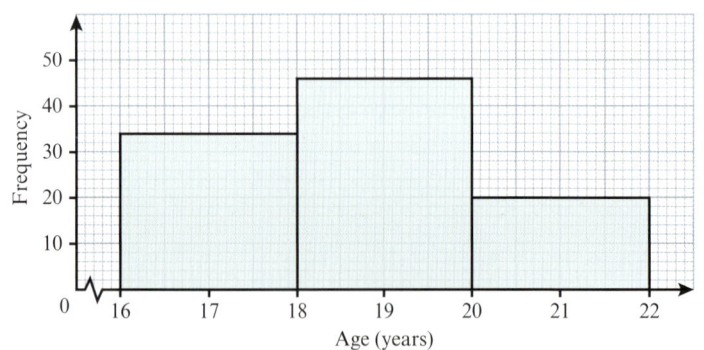

Chapter 1: Representation of data

The following table shows the areas of the columns and the **frequency** of each of the three classes presented in the diagram on the previous page.

	First	Second	Third
Area	2 × 34 = 68	2 × 46 = 92	2 × 20 = 40
Frequency	34	46	20

From this table we can see that the ratio of the column areas, 68 : 92 : 40, is exactly the same as the ratio of the frequencies, 34 : 46 : 20.

In a histogram, the area of a column represents the frequency of the corresponding class, so that the area must be proportional to the frequency.

We may see this written as 'area ∝ frequency'.

This also means that in every histogram, just as in the example above, the ratio of column areas is the same as the ratio of the frequencies, even if the classes do not have equal widths.

Also, there can be no gaps between the columns in a histogram because the upper boundary of one class is equal to the lower boundary of the neighbouring class. A gap can appear only when a class has zero frequency.

The axis showing the measurements is labelled as a continuous number line, and the width of each column is equal to the width of the class that it represents.

When we construct a histogram, since the classes may not have equal widths, the height of each column is no longer determined by the frequency alone, but must be calculated so that area ∝ frequency.

The vertical axis of the histogram is labelled **frequency density**, which measures frequency per standard interval. The simplest and most commonly used standard interval is 1 unit of measurement.

For example, if we require a column to represent 85 objects, each with $50 \leqslant \text{mass} < 60$ kg then an appropriate standard interval to use is 1 kilogram.

Height = $\dfrac{\text{area}}{\text{width}}$, so Frequency density = $\dfrac{85 \text{ objects}}{(60 - 50) \text{ kg}}$ = 8.5 objects per kilogram.

Alternatively, we could use a standard interval of 1 gram instead:

Frequency density = $\dfrac{85 \text{ objects}}{(60\,000 - 50\,000) \text{ g}}$ = 0.0085 objects per gram.

8.5 objects/kg and 0.0085 objects/g are equivalent densities.

> **TIP**
>
> The symbol ∝ means 'is proportional to'.

> **TIP**
>
> A standard interval other than 1 unit may be used, so the height of the column for these 85 objects could be labelled with any equivalent density such as: 17 objects per 2 kg or 3.4 objects per 400 g and so on.

KEY POINT 1.3

For a standard interval of 1 unit of measurement, Frequency density = $\dfrac{\text{class frequency}}{\text{class width}}$, which can be rearranged to give

Class frequency = class width × frequency density

In a histogram, we can see the relative frequencies of classes by comparing column areas, and we can make estimates by assuming that the values in each class are spread evenly over the whole **class interval**.

WORKED EXAMPLE 1.2

The masses, m kg, of 100 children are grouped into two classes, as shown in the table.

Mass (m kg)	$40 \leqslant m < 50$	$50 \leqslant m < 70$
No. children (f)	40	60

a Illustrate the data in a histogram.

b Estimate the number of children with masses between 45 and 63 kg.

Answer

a

Mass (m kg)	$40 \leqslant m < 50$	$50 \leqslant m < 70$
No. children (f)	40	60
Class width (kg)	10	20
Frequency density	$40 \div 10 = 4$	$60 \div 20 = 3$

Frequency density is calculated for the unequal-width intervals in the table. The masses are represented in the histogram, where frequency density measures *number of children per 1 kg* or simply *children per kg*.

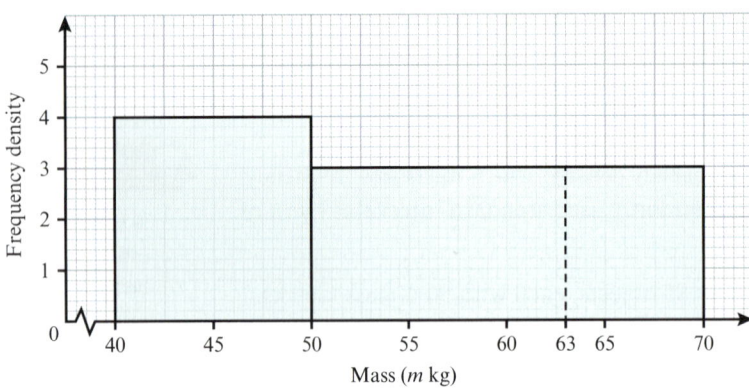

> **TIP**
>
> Column areas are equal to class frequencies. For example, the area of the first column is
> $(50 - 40)\,\text{kg} \times \dfrac{4 \text{ children}}{1 \text{ kg}}$
> = 40 children.

> **TIP**
>
> If we drew column heights of 8 and 6 instead of 4 and 3, then frequency density would measure *children per 2 kg*. The area of the first column would be
> $(50-40)\,\text{kg} \times \dfrac{8 \text{ children}}{2 \text{ kg}}$
> = 40 children.

b There are children with masses from 45 to 63 kg in both classes, so we must split this interval into two parts: 45–50 and 50–63.

$\dfrac{1}{2} \times 40 = 20$ children   The class 40–50 kg has a mid-value of 45 kg and a frequency of 40.

Frequency = width × frequency density

$= (63 - 50)\,\text{kg} \times \dfrac{3 \text{ children}}{1 \text{ kg}}$

$= 13 \times 3$ children

$= 39$ children

Our estimate for the interval 50–63 kg is equal to the area corresponding to this section of the second column.

Our estimate is $20 + 39 = 59$ children.   We add together the estimates for the two intervals.

Consider the times taken, to the nearest minute, for 36 athletes to complete a race, as given in the table below.

Time taken (min)	13	14–15	16–18
No. athletes (f)	4	14	18

Gaps of 1 minute appear between classes because the times are rounded.

Frequency densities are calculated in the following table.

Time taken (t min)	$12.5 \leq t < 13.5$	$13.5 \leq t < 15.5$	$15.5 \leq t < 18.5$
No. athletes (f)	4	14	18
Class width (min)	1	2	3
Frequency density	4 ÷ 1 = 4	14 ÷ 2 = 7	18 ÷ 3 = 6

This histogram represents the race times, where frequency density measures *athletes per minute*.

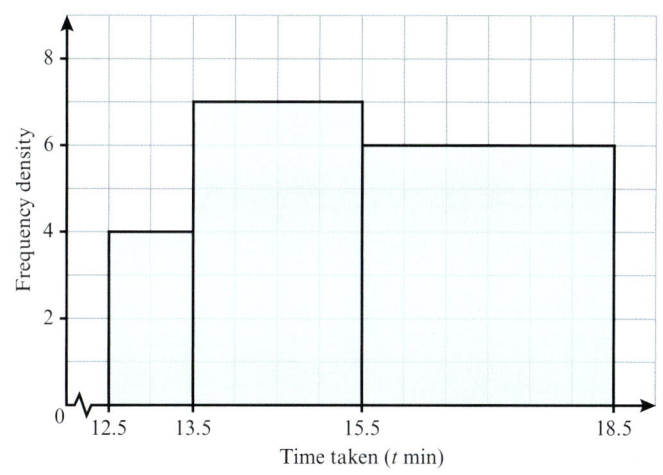

> **TIP**
> Use class boundaries (rather than rounded values) to find class widths, otherwise incorrect frequency densities will be obtained.

> **TIP**
> The class with the highest frequency does not necessarily have the highest frequency density.

> **TIP**
> Do think carefully about the scales you use when constructing a histogram or any other type of diagram. Sensible scales, such as 1 cm for 1, 5, 10, 20 or 50 units, allow you to read values with much greater accuracy than scales such as 1 cm for 3, 7 or 23 units. For similar reasons, try to use as much of the sheet of graph paper as possible, ensuring that the whole diagram will fit before you start to draw it.

WORKED EXAMPLE 1.3

Use the histogram of race times shown previously to estimate:
   a the number of athletes who took less than 13.0 minutes
   b the number of athletes who took between 14.5 and 17.5 minutes
   c the time taken to run the race by the slowest three athletes.

Answer

We can see that two blocks represent one athlete in the histogram. So, instead of calculating with frequency densities, we can simply count the number of blocks and divide by 2 to estimate the number of athletes involved.

a 4 ÷ 2 = 2 athletes .... There are four blocks to the left of 13.0 minutes.

b 38 ÷ 2 = 19 athletes .... There are 14 + 24 = 38 blocks between 14.5 and 17.5 minutes.

c Between 18.0 and 18.5 minutes. .... The slowest three athletes are represented by the six blocks to the right of 18.0 minutes.

Cambridge International AS & A Level Mathematics: Probability & Statistics 1

EXPLORE 1.1

Refer back to the table in Section 1.3 that shows the percentage scores of 100 students who took an examination.

Discuss what adjustments must be made so that the data can be represented in a histogram.

How could we make these adjustments and is there more than one way of doing this?

> **TIP**
>
> It is not acceptable to draw the axes or the columns of a histogram freehand. Always use a ruler!

EXERCISE 1B

1. In a particular city there are 51 buildings of historical interest. The following table presents the ages of these buildings, given to the nearest 50 years.

Age (years)	50–150	200–300	350–450	500–600
No. buildings (f)	15	18	12	6

   a   Write down the lower and upper boundary values of the class containing the greatest number of buildings.

   b   State the widths of the four class intervals.

   c   Illustrate the data in a histogram.

   d   Estimate the number of buildings that are between 250 and 400 years old.

2. The masses, m grams, of 690 medical samples are given in the following table.

Mass (m grams)	$4 \leqslant m < 12$	$12 \leqslant m < 24$	$24 \leqslant m < 28$
No. medical samples (f)	224	396	p

   a   Find the value of p that appears in the table.

   b   On graph paper, draw a histogram to represent the data.

   c   Calculate an estimate of the number of samples with masses between 8 and 18 grams.

3. The table below shows the heights, in metres, of 50 boys and of 50 girls.

Height (m)	1.2–	1.3–	1.6–	1.8–1.9
No. boys (f)	7	11	26	6
No. girls (f)	10	22	16	2

   a   How many children are between 1.3 and 1.6 metres tall?

   b   Draw a histogram to represent the heights of all the boys and girls together.

   c   Estimate the number of children whose heights are 1.7 metres or more.

4. The heights of 600 saplings are shown in the following table.

Height (cm)	0–	5–	15–	30–u
No. saplings (f)	64	232	240	64

   a   Suggest a suitable value for u, the upper boundary of the data.

   b   Illustrate the data in a histogram.

   c   Calculate an estimate of the number of saplings with heights that are:

   i   less than 25 cm   ii   between 7.5 and 19.5 cm.

5 Each of the 70 trainees at a secretarial college was asked to type a copy of a particular document. The times taken are shown, correct to the nearest 0.1 minutes, in the following table.

Time taken (min)	2.6–2.8	2.9–3.0	3.1–3.2	3.3–3.7
No. trainees (f)	15	25	20	10

a Explain why the interval for the first class has a width of 0.3 minutes.

b Represent the times taken in a histogram.

c Estimate, to the nearest second, the upper boundary of the times taken by the fastest 10 typists.

d It is given that 15 trainees took between 3.15 and b minutes. Calculate an estimate for the value of b when:

   i $b > 3.15$ ii $b < 3.15$.

6 A railway line monitored 15% of its August train journeys to find their departure delay times. The results are shown below. It is given that 24 of these journeys were delayed by less than 2 minutes.

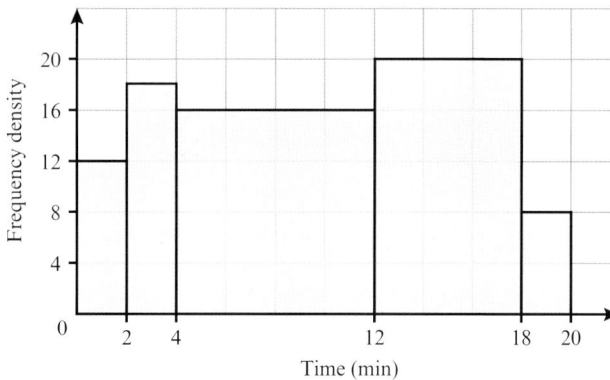

a How many journeys were monitored?

b Calculate an estimate of the number of these journeys that were delayed by:

   i 1 to 3 minutes   ii 10 to 15 minutes.

c Show that a total of 2160 journeys were provided in August.

d Calculate an estimate of the number of August journeys that were delayed by 3 to 7 minutes. State any assumptions that you make in your calculations.

7 A university investigated how much space on its computers' hard drives is used for data storage. The results are shown below. It is given that 40 hard drives use less than 20 GB for data storage.

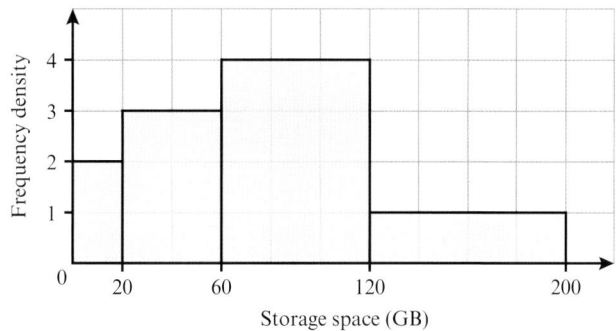

a Find the total number of hard drives represented.

b Calculate an estimate of the number of hard drives that use less than 50 GB.

c Estimate the value of k, if 25% of the hard drives use k GB or more.

8 The lengths of the 575 items in a candle maker's workshop are represented in the histogram.

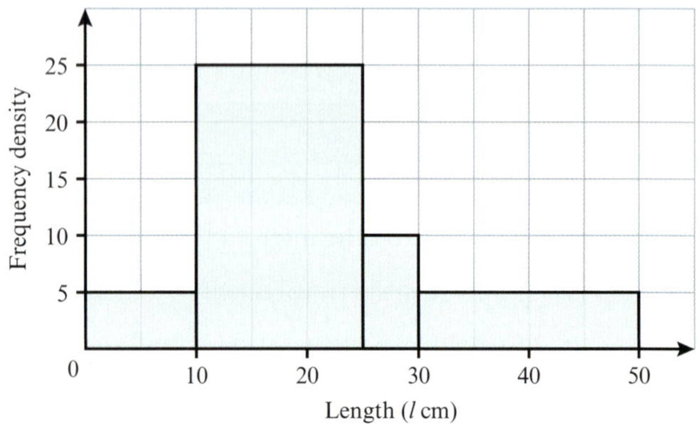

   a What proportion of the items are less than 25 cm long?

   b Estimate the number of items that are between 12.4 and 36.8 cm long.

   c The shortest 20% of the workshop's items are to be recycled. Calculate an estimate of the length of the shortest item that will not be recycled.

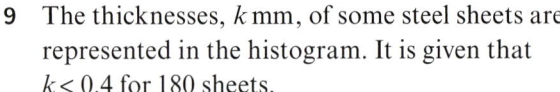

9 The thicknesses, k mm, of some steel sheets are represented in the histogram. It is given that $k < 0.4$ for 180 sheets.

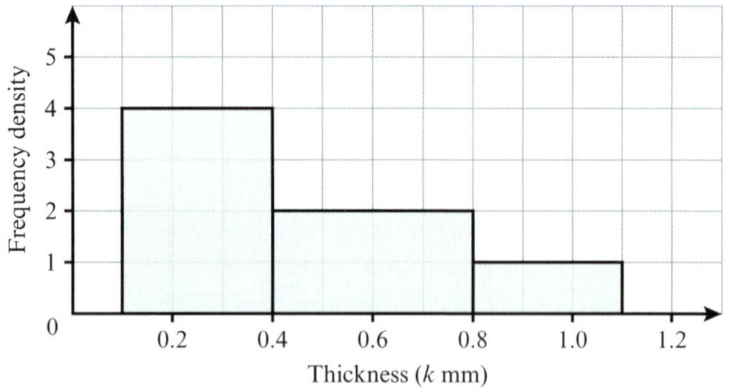

   a Find the ratio between the frequencies of the three classes. Give your answer in simplified form.

   b Find the value of n, given that frequency density measures sheets per n mm.

   c Calculate an estimate of the number of sheets for which:

      i $k < 0.5$ ii $0.75 \leqslant k < 0.94$.

   d The sheets are classified as thin, medium or thick in the ratio $1:3:1$.

      Estimate the thickness of a medium sheet, giving your answer in the form $a \leqslant k < b$. How accurate are your values for a and b?

10 The masses, in kilograms, of the animals treated at a veterinary clinic in the past year are illustrated in a histogram. The histogram has four columns of equal height. The following table shows the class intervals and the number of animals in two of the classes.

Mass (kg)	3–5	6–12	13–32	33–44
No. animals (f)	a	371	1060	b

   a Find the value of a and of b, and show that a total of 2226 animals were treated at the clinic.

   b Calculate an estimate of the lower boundary of the masses of the heaviest 50% of these animals.

11 The minimum daily temperature at a mountain village was recorded to the nearest $0.5\,°C$ on 200 consecutive days. The results are grouped into a frequency table and a histogram is drawn.

   The temperatures ranged from $0.5\,°C$ to $2\,°C$ on n days, and this class is represented by a column of height h cm.

   The temperatures ranged from -2.5 to $-0.5\,°C$ on d days. Find, in terms of n, h and d, the height of the column that represents these temperatures.

12 The frequency densities of the four classes in a histogram are in the ratio $4:3:2:1$. The frequencies of these classes are in the ratio $10:15:24:8$.

   Find the total width of the histogram, given that the narrowest class interval is represented by a column of width 3 cm.

 13 The percentage examination scores of 747 students are given in the following table.

Score (%)	p–50	51–70	71–80	81–q
No. students (f)	165	240	195	147

Given that the frequency densities of the four classes of percentage scores are in the ratio $5:8:13:7$, find the value of p and of q.

 DID YOU KNOW?

Bar charts first appeared in a book by the Scottish political economist William Playfair, entitled *The Commercial and Political Atlas* (London, 1786). His invention was adopted by many in the following years, including Florence Nightingale, who used bar charts in 1859 to compare mortality in the peacetime army with the mortality of civilians. This helped to convince the government to improve army hygiene.

In the past few decades histograms have played a very important role in image processing and computer vision. An image histogram acts as a graphical representation of the distribution of colour tones in a digital image. By making adjustments to the histogram, an image can be greatly enhanced. This has had great benefits in medicine, where scanned images are used to diagnose injury and illness.

FAST FORWARD

In Chapter 2, Section 2.3 and in Chapter 8, Section 8.1, we will see how a histogram or bar chart can be used to show the *shape* of a set of data, and how that shape provides information on average values.

1.4 Representation of continuous data: cumulative frequency graphs

A **cumulative frequency graph** can be used to represent continuous data. **Cumulative frequency** is the total frequency of all values less than a given value.

If we are given grouped data, we can construct the cumulative frequency diagram by plotting cumulative frequencies (abbreviated to *cf*) against upper class boundaries for all intervals. We can join the points consecutively with straight-line segments to give a cumulative frequency polygon or with a smooth curve to give a cumulative frequency curve.

For example, a set of data that includes 100 values below 7.5 and 200 values below 9.5 will have two of its points plotted at (7.5, 100) and at (9.5, 200).

We plot points at upper boundaries because we know the total frequencies up to these points are precise.

From a cumulative frequency graph we can estimate the number or proportion of values that lie above or below a given value, or between two values. There is no rule for deciding whether a polygon or curve is the best type of graph to draw. It is often difficult to fit a smooth curve through a set of plotted points. Also, it is unlikely that any two people will draw exactly the same curve. A polygon, however, is not subject to this uncertainty, as we know exactly where the line segments must be drawn.

 TIP

A common mistake is to plot points at class mid-values but the total frequency up to each mid-value is not precise – it is an estimate.

By drawing a cumulative frequency polygon, we are making exactly the same assumptions that we made when we used a histogram to calculate estimates. This means that estimates from a polygon should match exactly with estimates from a histogram.

> **REWIND**
>
> In Worked example 1.2, we made estimates from a histogram by assuming that the values in each class are spread evenly over the whole class interval.

WORKED EXAMPLE 1.4

The following table shows the lengths of 80 leaves from a particular tree, given to the nearest centimetre.

Lengths (cm)	1–2	3–4	5–7	8–9	10–11
No. leaves (f)	8	20	38	10	4

Draw a cumulative frequency curve and a cumulative frequency polygon. Use each of these to estimate:

a the number of leaves that are less than 3.7 cm long

b the lower boundary of the lengths of the longest 22 leaves.

Answer

Lengths (l cm)	Addition of frequencies	No. leaves (cf)
$l < 0.5$		0
$l < 2.5$	$0 + 8$	8
$l < 4.5$	$0 + 8 + 20$	28
$l < 7.5$	$0 + 8 + 20 + 38$	66
$l < 9.5$	$0 + 8 + 20 + 38 + 10$	76
$l < 11.5$	$0 + 8 + 20 + 38 + 10 + 4$	80

Before the diagrams can be drawn, we must organise the given data to show upper class boundaries and cumulative frequencies.

76 can also be calculated as 66 + 10, using the previous cf value.

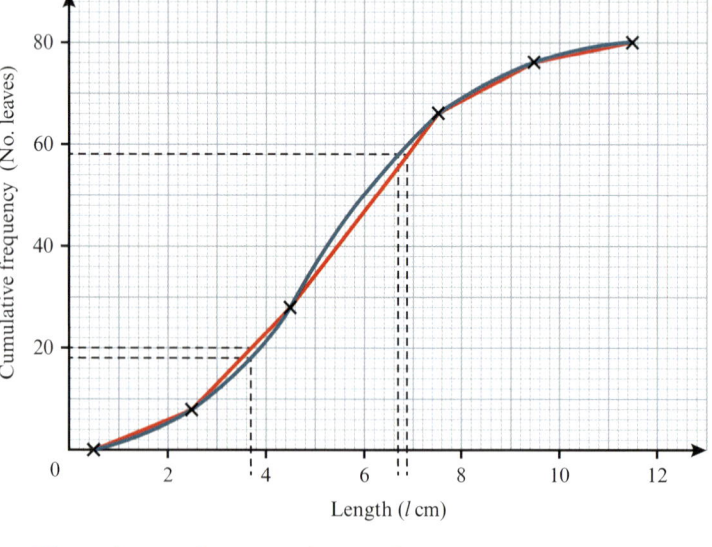

We plot points at (0.5, 0), (2.5, 8), (4.5, 28), (7.5, 66), (9.5, 76) and (11.5, 80). We then join them in order with ruled lines and also with a smooth curve, to give the two types of graph.

a The polygon gives an estimate of 20 leaves.

   The curve gives an estimate of 18 leaves.

b The polygon gives an estimate of 6.9 cm.

   The curve gives an estimate of 6.7 cm.

> **TIP**
>
> - Do include the lowest class boundary, which has a cumulative frequency of 0, and plot this point on your graph.
> - The dotted lines show the workings for parts **a** and **b**.
> - When constructing a cumulative frequency graph, you are advised to use sensible scales that allow you to plot and read values accurately.
> - Estimates from a polygon and a curve will not be the same, as they coincide only at the plotted points.
> - Note that all of these answers are only estimates, as we do not know the exact shape of the cumulative frequency graph between the plotted points.

Chapter 1: Representation of data

EXPLORE 1.2

Four histograms (1–4) that represent four different sets of data with equal-width intervals are shown.

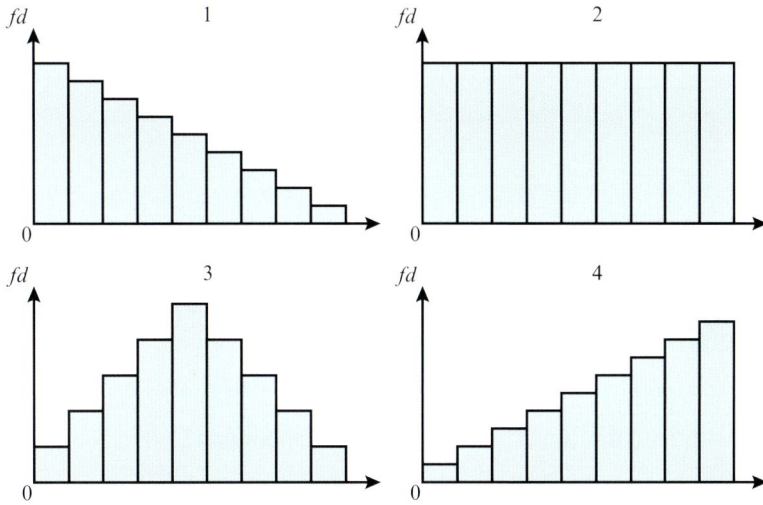

a Which of the following cumulative frequency graphs (A–D) could represent the same set of data as each of the histograms (1–4) above?

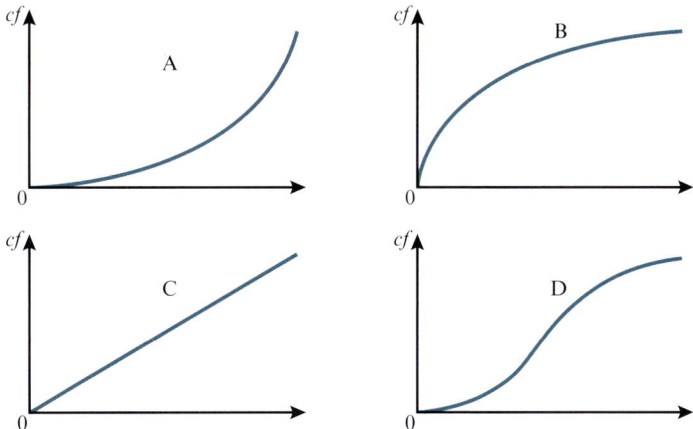

How do column heights affect the shape of a cumulative frequency graph?

b Why could a cumulative frequency graph never look like the sketch shown below?

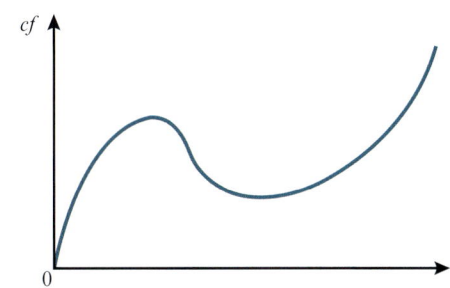

FAST FORWARD

We will use cumulative frequency graphs to estimate the median, quartiles and percentiles of a set of data in Chapter 2, Section 2.3 and in Chapter 3, Section 3.2.

WEB LINK

You can investigate the relationship between histograms and cumulative frequency graphs by visiting the Cumulative Frequency Properties resource on the Geogebra website (www.geogebra.org).

EXERCISE 1C

1. The reaction times, t seconds, of 66 participants were measured in an experiment and presented below.

Time (t seconds)	No. participants (cf)
$t < 1.5$	0
$t < 3.0$	3
$t < 4.5$	8
$t < 6.5$	32
$t < 8.5$	54
$t < 11.0$	62
$t < 13.0$	66

   a Draw a cumulative frequency polygon to represent the data.

   b Use your graph to estimate:

      i the number of participants with reaction times between 5.5 and 7.5 seconds

      ii the lower boundary of the slowest 20 reaction times.

2. The following table shows the widths of the 70 books in one section of a library, given to the nearest centimetre.

Width (cm)	10–14	15–19	20–29	30–39	40–44
No. books (f)	3	13	25	24	5

   a Given that the upper boundary of the first class is 14.5 cm, write down the upper boundary of the second class.

   b Draw up a cumulative frequency table for the data and construct a cumulative frequency graph.

   c Use your graph to estimate:

      i the number of books that have widths of less than 27 cm

      ii the widths of the widest 20 books.

3. Measurements of the distances, x mm, between two moving parts inside car engines were recorded and are summarised in the following table. There were 156 engines of type A and 156 engines of type B.

Distance (x mm)	$x < 0.10$	$x < 0.35$	$x < 0.60$	$x < 0.85$	$x < 1.20$
Engine A (cf)	0	16	84	144	156
Engine B (cf)	0	8	62	120	156

   a Draw and label two cumulative frequency curves on the same axes.

   b Use your graphs to estimate:

      i the number of engines of each type with measurements between 0.30 and 0.70 mm

      ii the total number of engines with measurements that were less than 0.55 mm.

   c Both types of engine must be repaired if the distance between these moving parts is more than a certain fixed amount. Given that 16 type A engines need repairing, estimate the number of type B engines that need repairing.

4 The diameters, d cm, of 60 cylindrical electronic components are represented in the following cumulative frequency graph.

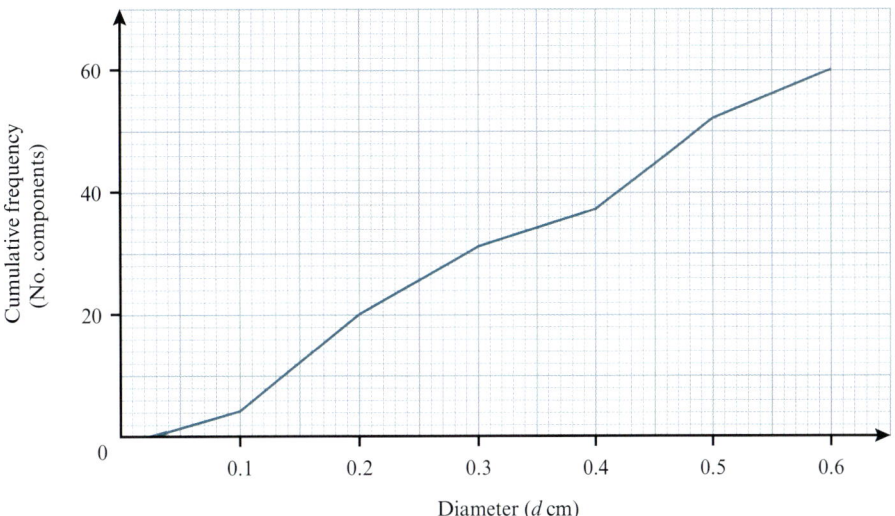

a Find the number of components such that $0.2 \leqslant d < 0.4$, and explain why your answer is not an estimate.

b Estimate the number of components that have:

   i a diameter of less than 0.15 cm   ii a radius of 0.16 cm or more.

c Estimate the value of k, given that 20% of the components have diameters of k mm or more.

d Give the reason why 0.1–0.2 cm is the modal class.

5 The following cumulative frequency graph shows the masses, m grams, of 152 uncut diamonds.

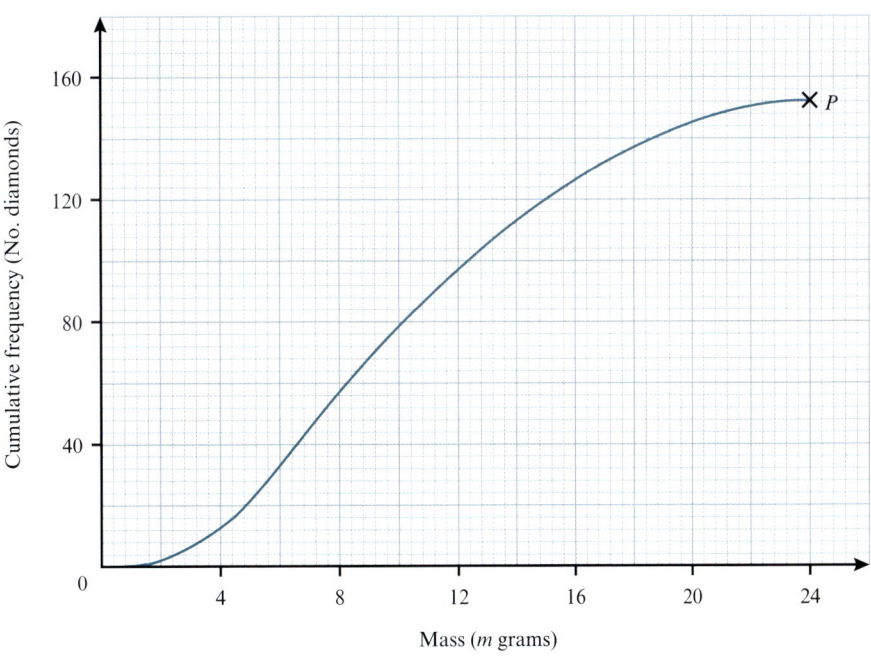

a Estimate the number of uncut diamonds with masses such that:

   i $9 \leqslant m < 17$ ii $7.2 \leqslant m < 15.6$.

b The lightest 40 diamonds are classified as small. The heaviest 40 diamonds are classified as large. Estimate the difference between the mass of the heaviest small diamond and the lightest large diamond.

c The point marked at $P(24, 152)$ on the graph indicates that the 152 uncut diamonds all have masses of less than 24 grams.

Each diamond is now cut into two parts of equal mass. Assuming that there is no wastage of material, write down the coordinates of the point corresponding to P on a cumulative frequency graph representing the masses of these cut diamonds.

6 The densities, d g/cm³, of 125 chemical compounds are given in the following table.

Density (d g/cm³)	$d < 1.30$	$d < 1.38$	$d < 1.42$	$d < 1.58$	$d < 1.70$
No. compounds (cf)	32	77	92	107	125

Find the frequencies a, b, c and d given in the table below.

Density (d g/cm³)	0–	1.30–	1.38–	1.42–1.70
No. compounds (f)	a	b	c	d

7 The daily journey times for 80 bank staff to get to work are given in the following table.

Time (t min)	$t < 10$	$t < 15$	$t < 20$	$t < 25$	$t < 30$	$t < 45$	$t < 60$
No. staff (cf)	3	11	24	56	68	76	80

a How many staff take between 15 and 45 minutes to get to work?

b Find the exact number of staff who take $\dfrac{x+y}{2}$ minutes or more to get to work, given that 85% of the staff take less than x minutes and that 70% of the staff take y minutes or more.

8 A fashion company selected 100 12-year-old boys and 100 12-year-old girls to audition as models. The heights, h cm, of the selected children are represented in the following graph.

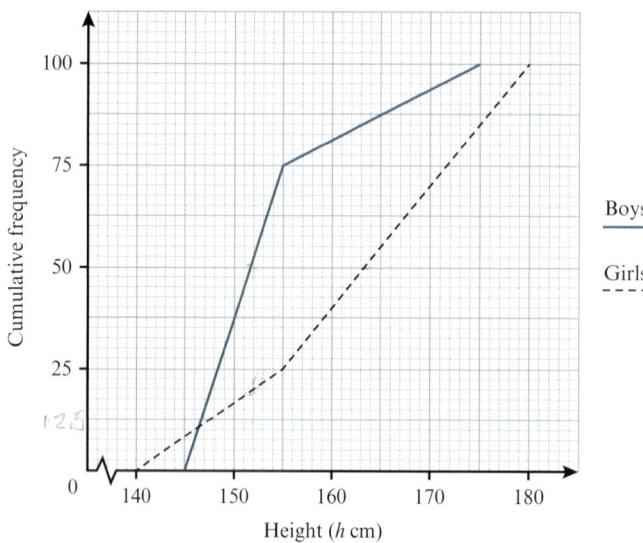

a What features of the data suggest that the children were not selected at random?

b Estimate the number of girls who are taller than the shortest 50 boys.

c What is the significance of the value of h where the graphs intersect?

d The shortest 75 boys and tallest 75 girls were recalled for a second audition. On a cumulative frequency graph, show the heights of the children who were not recalled.

9 The following table shows the ages of the students currently at a university, given by percentage. Ages are rounded down to the number of whole years.

Age (years)	<18	18–19	20–21	22–24	25–28	29–35	36–44
Students (%)	0	27	51	11	5	4	2

  a Represent the data in a percentage cumulative frequency polygon.

  b The oldest 8% of these students qualify as 'mature'. Use your polygon to estimate the minimum age requirement for a student to be considered mature. Give your answer to the nearest month.

  c Of the 324 students who are 18–19 years old, 54 are not expected to find employment within 3 months of finishing their course.

   i Calculate an estimate of the number of current students who are expected to find employment within 3 months of finishing their course.

   ii What assumptions must be made to justify your calculations in part c i? Are these assumptions reasonable? Do you expect your estimate to be an overestimate or an underestimate?

10 The distances, in km, that 80 new cars can travel on 1 litre of fuel are shown in the table.

Distance (km)	4.4–	6.6–	8.8–	12.1–	15.4–18.7
No. cars (f)	5	7	52	12	4

These distances are 10% greater than the distances the cars will be able to travel after they have covered more than 100 000 km.

Estimate how many of the cars can travel 10.5 km or more on 1 litre of fuel when new, but not after they have covered more than 100 000 km.

11 A small company produces cylindrical wooden pegs for making garden chairs. The lengths and diameters of the 242 pegs produced yesterday have been measured independently by two employees, and their results are given in the following table.

Length (l cm)	$l < 1.0$	$l < 2.0$	$l < 2.5$	$l < 3.0$	$l < 3.5$	$l < 4.0$	$l < 4.5$
No. pegs (cf)	0	0	8	40	110	216	242

Diameter (d cm)	$d < 1.0$	$d < 1.5$	$d < 2.0$	$d < 2.5$	$d < 3.0$
No. pegs (cf)	0	60	182	222	242

  a On the same axes, draw two cumulative frequency graphs: one for lengths and one for diameters.

  b Correct to the nearest millimetre, the lengths and diameters of n of these pegs are equal. Find the least and greatest possible value of n.

  c A peg is acceptable for use when it satisfies both $l \geq 2.8$ and $d < 2.2$.

   Explain why you cannot obtain from your graphs an accurate estimate of the number of these 242 pegs that are acceptable. Suggest what the company could do differently so that an accurate estimate of the proportion of acceptable pegs could be obtained.

EXPLORE 1.3

The following table shows data on the masses, m grams, of 150 objects.

Mass (m g)	$m < 0$	$m < 12$	$m < 30$	$m < 53$	$m < 70$	$m < 80$
No. objects (cf)	0	24	60	106	138	150

By drawing a cumulative frequency polygon, the following estimates will be obtained:

a Number of objects with masses less than 20 g = 40 objects.

b Arranged in ascending order, the mass of the 100th object = 50 g.

However, we can calculate these estimates from the information given in the table without drawing the polygon.

Investigate the possible methods that we can use to calculate these two estimates.

> **TIP**
>
> The six plotted points, whose coordinates you know, are joined by straight lines.

1.5 Comparing different data representations

Pictograms, vertical line graphs, bar charts and pie charts are useful ways of displaying qualitative data and ungrouped quantitative data, and people generally find them easy to understand. Nevertheless, it may be of benefit to group a set of raw data so that we can see how the values are distributed. Knowing the proportion of small, medium and large values, for example, may prove to be useful. For small datasets we can do this by constructing stem-and-leaf diagrams, which have the advantage that raw values can still be seen after grouping.

In large datasets individual values lose their significance and a picture of the whole is more informative. We can use frequency tables to make a compact summary by grouping but most people find the information easier to grasp when it is shown in a graphical format, which allows absolute, relative or cumulative frequencies to be seen. Although some data are lost by grouping, histograms and cumulative frequency graphs have the advantage that data can be grouped into classes of any and varied widths.

The choice of which representation to use will depend on the type and quantity of data, the audience and the objectives behind making the representation. Most importantly, the representation must show the data clearly and should not be misleading in any way.

The following chart is a guide to some of the most commonly used methods of data representation.

> **FAST FORWARD**
>
> In Chapter 2 and Chapter 3, we will see how grouping effects the methods we use to find measures of central tendency and measures of variation.

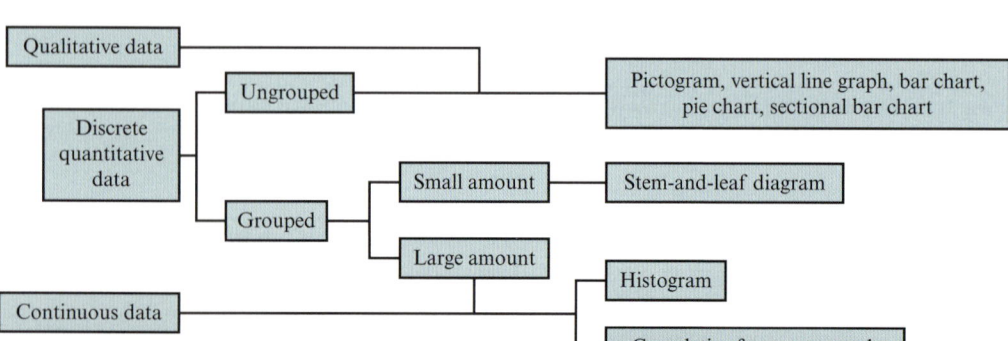

EXERCISE 1D

1. Jamila noted each student's answer when her year group was asked to name their favourite colour.

   a. List the methods of representation that would be suitable for displaying Jamila's data.

   b. Jamila wishes to emphasise that the favourite colour of exactly three-quarters of the students is blue. Which type of representation from your list do you think would be the most effective for Jamila to use? Explain why you have chosen this particular type of representation.

2. A large number of chickens' eggs are individually weighed. The masses are grouped into nine classes, each of width 2 grams, from 48 to 66 g.

   Name a type of representation in which the fact could be seen that the majority of the eggs have masses from 54 to 60 g. Explain how the representation would show this.

3. Boxes of floor tiles are to be offered for sale at a special price of $75. The boxes claim to contain at least 100 tiles each.

   a. Why would it be preferable to use a stem-and-leaf diagram rather than a bar chart to represent the numbers of tiles, which are 112, 116, 107, 108, 121, 104, 111, 106, 105 and 110?

   b. How may the seller benefit if the numbers 12, 16, 7, 8, 21, 4, 11, 6, 5 and 10 are used to draw the stem-and-leaf diagram instead of the actual numbers of tiles?

4. A charity group's target is to raise a certain amount of money in a year. At the end of the first month the group raised 36% of the target amount, and at the end of each subsequent month they manage to raise exactly half of the amount outstanding.

   a. How many months will it take the group to raise 99% of the money?

   b. Name a type of representation that will show that the group fails to reach its target by the end of the year. Explain how this fact would be shown in the representation.

5. University students measured the heights of the 54 trees in the grounds of a primary school. As part of a talk on conservation at a school assembly, the students have decided to present their data using one of the following diagrams.

   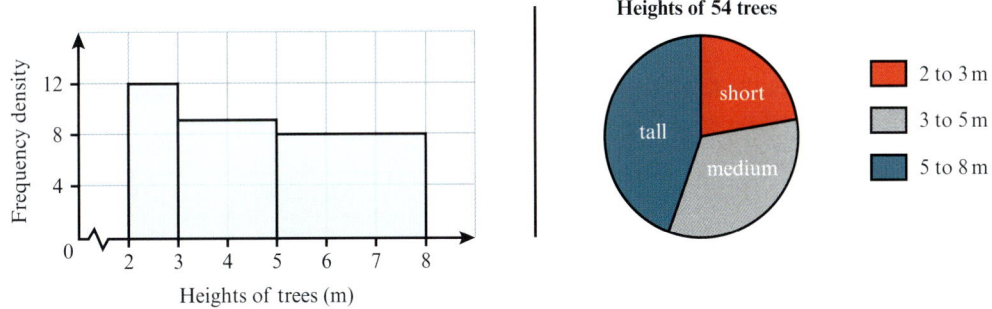

   a. Give one disadvantage of using each of the representations shown.

   b. Name and describe a different type of representation that would be appropriate for the audience, and that has none of the disadvantages given in part **a**.

6 The percentage scores of 40 candidates who took a Health and Safety test are given:

77 44 65 84 52 60 35 83 68 66 50 68 65 57 60 50 93 38 46 55

45 69 61 64 40 66 91 59 61 74 70 75 42 65 85 63 73 84 68 30

a Construct a frequency table by grouping the data into seven classes with equal-width intervals, where the first class is 30–39. Label this as Table 1.

b It is proposed that each of the 40 candidates is awarded one of three grades, A, B or C. Construct a new frequency table that matches with this proposal. Label this as Table 2.

c A student plans to display all three versions of the data (i.e. the raw data, the data in Table 1 and the data in Table 2) in separate stem-and-leaf diagrams.

For which version(s) of the data would this not be appropriate? Suggest an alternative type of representation in each case.

M 7 Last year Tom renovated an old building during which he worked for at least 9 hours each week. By plotting four points in a graph, he has represented the time he spent working.

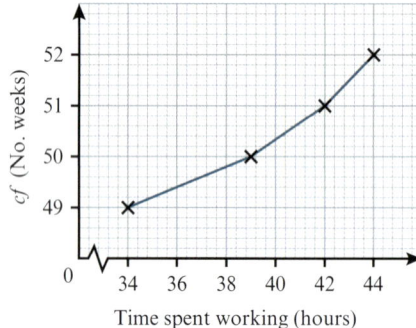

a What can you say about the time that Tom spent working on the basis of this graph?

b Explain why Tom's graph might be considered to be misleading.

c Name the different types of representation that are suitable for displaying the amount of time that Tom worked each week throughout the year.

Consider the benefits of each type of representation and then fully describe (but do not draw) the one you believe to be the most suitable.

PS 8 The following table shows the focal lengths, l mm, of the 84 zoom lenses sold by a shop. For example, there are 18 zoom lenses that can be set to any focal length between 24 and 50 mm.

Focal length (l mm)	24–50	50–108	100–200	150–300	250–400
No. lenses (f)	18	30	18	12	6

a What feature of the data does not allow them to be displayed in a histogram?

b What type of diagram could you use to illustrate the data? Explain clearly how you would do this.

Chapter 1: Representation of data

 9   The following table shows estimates, in hundred thousands, of the number of people living in poverty and the populations, in millions, of the countries where they live. (Source: *World Bank 2011/12*)

Country	Chile	Sri Lanka	Malaysia	Georgia	Mongolia
No. living in poverty ($\times 10^5$)	24.8	18.4	10.8	7.91	7.50
Population ($\times 10^6$)	17.25	20.65	28.33	4.47	2.74

Represent, in a single diagram, the actual numbers and the relative poverty that exists in these countries.

In what way do the two sets of data in your representation give very different pictures of the poverty levels that exist? Which is the better representation to use and why?

EXPLORE 1.4

Past, present and predicted world population figures by age group, sex and other categories can be found on government census websites.

You may be interested, for example, in the population changes for your own age group during your lifetime. This is something that can be represented in a diagram, either manually, using spreadsheet software or an application such as GeoGebra, and for which you may like to try making predictions by looking for trends shown in the raw data or in any diagrams you create.

Checklist of learning and understanding

- Non-numerical data are called qualitative or categorical data.
- Numerical data are called quantitative data, and are either discrete or continuous.
  - Discrete data can take only certain values.
  - Continuous data can take any value, possibly within a limited range.
- Data in a stem-and-leaf diagram are ordered in rows with intervals of equal width.
- In a histogram, column area $\propto$ frequency, and the vertical axis is labelled frequency density.
  - Frequency density = $\dfrac{\text{class frequency}}{\text{class width}}$ and Class frequency = class width $\times$ frequency density.
- In a cumulative frequency graph, points are plotted at class upper boundaries.

Cambridge International AS & A Level Mathematics: Probability & Statistics 1

END-OF-CHAPTER REVIEW EXERCISE 1

 1 The weights of 220 sausages are summarised in the following table.

Weight (grams)	< 20	< 30	< 40	< 45	< 50	< 60	< 70
Cumulative frequency	0	20	50	100	160	210	220

  i State how many sausages weighed between 50g and 60g. [1]

  ii On graph paper, draw a histogram to represent the weights of the sausages. [4]

  Cambridge International AS & A Level Mathematics 9709 Paper 62 Q4 November 2011 [Adapted]

2 The lengths of children's feet, in centimetres, are classified as 14–16, 17–19, 20–22 and so on. State the lower class boundary, the class width and the class mid-value for the lengths given as 17–19. [2]

3 The capacities of ten engines, in litres, are given rounded to 2 decimal places, as follows:
1.86, 2.07, 1.74, 1.08, 1.99, 1.96, 1.83, 1.45, 1.28 and 2.19.

These capacities are to be grouped in three classes as 1.0–1.4, 1.5–2.0 and 2.1–2.2.

  a Find the frequency of the class 1.5–2.0 litres. [1]

  b Write down two words that describe the type of data given about the engines. [2]

4 Over a 10-day period, Alina recorded the number of text messages she received each day. The following stem-and-leaf diagram shows her results.

```
0 | 9            Key: 1 | 5
1 | 0 3 4 4     represents 15
1 | 5 5 5 6 8   messages
```

  a On how many days did she receive more than 10 but not more than 15 messages? [1]

  b How many more rows would need to be added to the stem-and-leaf diagram if Alina included data for two extra days on which she received 4 and 36 messages? Explain your answer. [2]

5 The following stem-and-leaf diagram shows eight randomly selected numbers between 2 and 4.

```
2 | 1 3 3 a     Key: 2 | 1
3 | b 5 6 7     represents 2.1
```

Given that $a - b = 7$ and that the sum of the eight numbers correct to the nearest integer is 24, find the value of a and of b. [3]

6 Eighty people downloaded a particular application and recorded the time taken for the download to complete. The times are given in the following table.

Download time (min)	< 3	< 5	< 6	< 10
No. downloads (*cf*)	6	18	66	80

  a Find the number of downloads that completed in 5 to 6 minutes. [1]

  b On a histogram, the download times from 5 to 6 minutes are represented by a column of height 9.6cm. Find the height of the column that represents the download times of 6 to 10 minutes. [2]

 7 A histogram is drawn with three columns whose widths are in the ratio 1 : 2 : 4. The frequency densities of these classes are in the ratio 16 : 12 : 3, respectively.

  a Given that the total frequency of the data is 390, find the frequency of each class. [3]

  b The classes with the two highest frequencies are to be merged and a new histogram drawn. Given that the height of the column representing the merged classes is to be 30cm, find the correct height for the remaining column. [3]

  c Explain what problems you would encounter if asked to construct a histogram in which the classes with the two lowest frequencies are to be merged. [1]

Chapter 1: Representation of data

8 The histograms below illustrate the number of hours of sunshine during August in two regions, A and B. Neither region had more than 8 hours of sunshine per day.

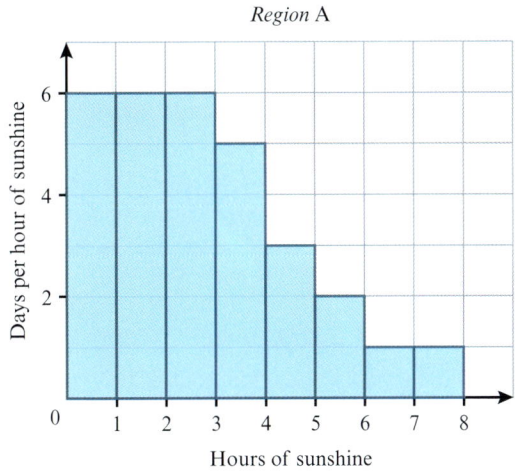
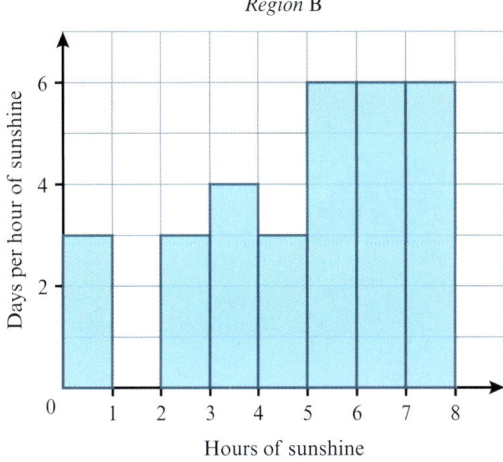

a Explain how you know that some information for one of the regions has been omitted. [2]

b After studying the histograms, two students make the following statements.

  • Bindu: There was more sunshine in region A than in region B during the first 2 weeks of August.
  • Janet: In August there was less sunshine in region A than in region B.

Discuss these statements and decide whether or not you agree with each of them.

In each case, explain your reasoning. [3]

9 A hotel has 90 rooms. The table summarises information about the number of rooms occupied each day for a period of 200 days.

Number of rooms occupied	1–20	21–40	41–50	51–60	61–70	71–90
Frequency	10	32	62	50	28	18

i Draw a cumulative frequency graph on graph paper to illustrate this information. [4]

ii Estimate the number of days when more than 30 rooms were occupied. [2]

iii On 75% of the days at most n rooms were occupied. Estimate the value of n. [2]

Cambridge International AS & A Level Mathematics 9709 Paper 62 Q5 June 2011

Chapter 2
Measures of central tendency

In this chapter you will learn how to:

- find and use different measures of central tendency
- calculate and use the mean of a set of data (including grouped data) either from the data itself or from a given total Σx or a coded total $\Sigma(x - b)$ and use such totals in solving problems that may involve up to two datasets.

Chapter 2: Measures of central tendency

PREREQUISITE KNOWLEDGE

Where it comes from	What you should be able to do	Check your skills
IGCSE / O Level Mathematics	Calculate the mean, median and mode for individual and discrete data.	1 Find the mean, median and mode of the numbers 7.3, 3.9, 1.3, 6.6, 9.2, 4.7, 3.9 and 3.1.
	Use a calculator efficiently and apply appropriate checks of accuracy.	2 Use a calculator to evaluate $\dfrac{6 \times 1.7 + 8 \times 1.9 + 11 \times 2.1}{6 + 8 + 11}$ and then check that your answer is reasonable.

Three types of average

There are three measures of central tendency that are commonly used to describe the *average* value of a set of data. These are the **mode**, the **mean** and the **median**.

- The mode is the most commonly occurring value.
- The mean is calculated by dividing the sum of the values by the number of values.
- The median is the value in the middle of an ordered set of data.

We use an average to **summarise** the values in a set of data. As a representative value, it should be fairly central to, and typical of, the values that it represents.

If we investigate the annual incomes of all the people in a region, then a single value (i.e. an average income) would be a convenient number to represent our findings. However, choosing which average to use is something that needs to be thought about, as one measure may be more appropriate to use than the others.

Deciding which measure to use depends on many factors. Although the mean is the most familiar average, a shoemaker would prefer to know which shoe size is the most popular (i.e. the mode). A farmer may find the median number of eggs laid by their chickens to be the most useful because they could use it to identify which chickens are profitable and which are not. As for the average income in our chosen region, we must also consider whether to calculate an average for the workers and managers together or separately; and, if separately, then we need to decide who fits into which category.

EXPLORE 2.1

Various sources tell us that the *average* person:

- laughs 10 times a day
- falls asleep in 7 minutes
- sheds 0.7 kg of skin each year
- grows 944 km of hair in a lifetime
- produces a sneeze that travels at 160 km/h
- has over 97 000 km of blood vessels in their body
- has a vocabulary between 5000 and 6000 words.

The average adult male is 172.5 cm tall and weighs 80 kg.

The average adult female is 159 cm tall and weighs 68 kg.

What might each of these statements mean and how might they have been determined?

Have you ever met such an average person? How could this information be useful?

You can find a variety of continuously updated figures that yield interesting averages at http://www.worldometers.info/.

2.1 The mode and the modal class

As you will recall, a set of data may have more than one mode or no mode at all.

The following table shows the scores on 25 rolls of a die, where 2 is the mode because it has the highest frequency.

Score on die	1	2	3	4	5	6
Frequency (f)	5	6	5	3	2	4

In a set of grouped data in which raw values cannot be seen, we can find the **modal class**, which is the class with the highest frequency density.

> **REWIND**
>
> We saw how to calculate frequency density in Chapter 1, Section 1.3.

> **KEY POINT 2.1**
>
> In histograms, the modal class has the greatest column height. If there is no modal class then all classes have the same frequency density.

WORKED EXAMPLE 2.1

Find the modal class of the 270 pencil lengths, given to the nearest centimetre in the following table.

Length (x cm)	No. pencils (f)
4–7	100
8–10	90
11–12	80

Answer

Length (x cm)	No. pencils (f)	Width (cm)	Frequency density
$3.5 \leq x < 7.5$	100	4	$100 \div 4 = 25$
$7.5 \leq x < 10.5$	90	3	$90 \div 3 = 30$
$10.5 \leq x < 12.5$	80	2	$80 \div 2 = 40$

Class boundaries, class widths and frequency density calculations are shown in the table.

Although the histogram shown is not needed to answer this question, it is useful to see that, in this case, the modal class is the one with the tallest column and the greatest frequency density, even though it has the lowest frequency.

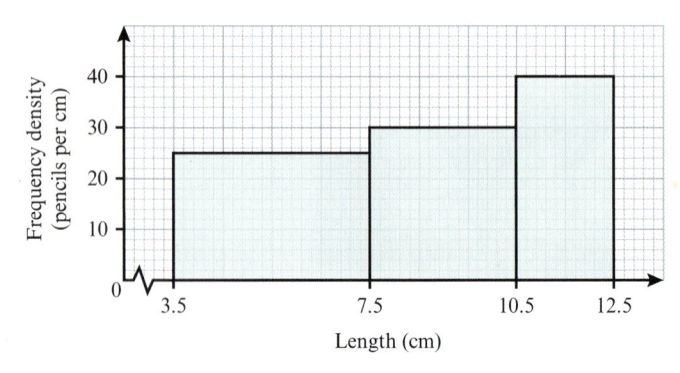

The modal class is 11–12 cm (or, more accurately, $10.5 \leq x < 12.5$ cm).

> **TIP**
>
> The modal class does not contain the most pencils but it does contain the greatest number of *pencils per centimetre*.

Chapter 2: Measures of central tendency

WORKED EXAMPLE 2.2

Two classes of data have interval widths in the ratio 3:2. Given that there is no modal class and that the frequency of the first class is 48, find the frequency of the second class.

Answer

Let the frequency of the second class be x.
$48 : x = 3 : 2$

$$\frac{x}{48} = \frac{2}{3}$$

$x = 32$

The second class has a frequency of 32.

'No modal class' means that the frequency densities of the two classes are equal, so class frequencies are in the same ratio as interval widths.

Or, let the frequency of the second class be x.

$$\frac{x}{2} = \frac{48}{3}$$

$x = 32$

Alternatively, frequencies are proportional to interval widths.

TIP

In the special case where all classes have equal widths, frequency densities are proportional to frequencies, so the modal class is the class with the highest frequency.

EXERCISE 2A

1. Find the mode(s) of the following sets of numbers.

    a 12, 15, 11, 7, 4, 10, 32, 14, 6, 13, 19, 3

    b 19, 21, 23, 16, 35, 8, 21, 16, 13, 17, 12, 19, 14, 9

2. Which of the eleven words in this sentence is the mode?

3. Identify the mode of x and of y in the following tables.

x	4	5	6	7	8
f	1	5	5	6	4

y	−4	−3	−2	−1	0
f	27	28	29	27	25

4. Find the modal class for x and for y in the following tables.

x	0–	4–	14–20
f	5	9	8

y	3–6	7–11	12–20
f	66	80	134

5. A small company sells glass, which it cuts to size to fit into window frames. How could the company benefit from knowing the modal size of glass its customers purchase?

6. Four classes of continuous data are recorded as 1–7, 8–16, 17–20 and 21–25. The class 1–7 has a frequency of 84 and there is no modal class. Find the total frequency of the other three classes.

PS 7. Data about the times, in seconds, taken to run 100 metres by n adults are given in the following table.

Time (x s)	$13.6 \leqslant x < 15.4$	$15.4 \leqslant x < 17.4$	$17.4 \leqslant x < 19.8$
No. adults (f)	a	b	27

By first investigating the possible values of a and of b, find the largest possible value of n, given that the modal class contains the slowest runners.

PS 8 Three classes of continuous data are given as $0 \leq x < 4$, $4 \leq x < 10$ and $10 \leq x < 18$. The frequency densities of the classes $0 \leq x < 4$ and $10 \leq x < 18$ are in the ratio $4:3$ and the total frequency of these two classes is 120. Find the least possible frequency of the modal class, given that the modal class is $4 \leq x < 10$.

2.2 The mean

The mean is referred to more precisely as the arithmetic mean and it is the most commonly known average. The sum of a set of data values can be found from the mean. Suppose, for example, that 12 values have a mean of 7.5:

$$\text{Mean} = \frac{\text{sum of values}}{\text{number of values}}, \text{ so } 7.5 = \frac{\text{sum of values}}{12} \text{ and sum of values} = 7.5 \times 12 = 90.$$

You will soon be performing calculations involving the mean, so here we introduce notation that is used in place of the word definition used above.

We use the upper case Greek letter 'sigma', written Σ, to represent 'sum' and $\bar{x}$ to represent the mean, where x represents our data values.

The notation used for ungrouped and for grouped data are shown on separate rows in the following table.

	Sum	Data values	Frequency of data values	Number of data values	Sum of data values	Mean
Ungrouped	Σ	x	–	n	Σx	$\bar{x} = \frac{\Sigma x}{n}$
Grouped	Σ	x	f	Σf	Σxf or Σfx	$\bar{x} = \frac{\Sigma xf}{\Sigma f}$

> **TIP**
> $n = \Sigma f$ for grouped data.

WORKED EXAMPLE 2.3

Five labourers, whose mean mass is 70.2 kg, wish to go to the top of a building in a lift with some cement. Find the greatest mass of cement they can take if the lift has a maximum weight allowance of 500 kg.

Answer

$\Sigma x = \bar{x} \times n$
$= 70.2 \times 5$
$= 351 \text{ kg}$

We first rearrange the formula $\bar{x} = \frac{\Sigma x}{n}$ to find the sum of the labourers' masses.

$351 + y = 500$
$y = 149$

We now form an equation using y to represent the greatest possible mass of cement.

The greatest mass of cement they can take is 149 kg.

> **TIP**
> $\Sigma xf = \Sigma fx$ indicates the sum of the products of each value and its frequency. For example, the sum of five 10s and six 20s is $(10 \times 5) + (20 \times 6) = (5 \times 10) + (6 \times 20) = 170$.

> **KEY POINT 2.2**
> For ungrouped data, the mean is $\bar{x} = \frac{\Sigma x}{n}$.
> For grouped data, $\bar{x} = \frac{\Sigma xf}{\Sigma f}$ or $\frac{\Sigma fx}{\Sigma f}$.

Chapter 2: Measures of central tendency

WORKED EXAMPLE 2.4

Find the mean of the 40 values of x, given in the following table.

x	31	32	33	34	35
f	5	7	9	8	11

Answer

x	31	32	33	34	35	
f	5	7	9	8	11	$\Sigma f = 40$
xf	155	224	297	272	385	$\Sigma xf = 1333$

We find the sum of the 40 values by adding together the products of each value of x and its frequency. This is done in the row headed xf in the table. The 40 values of x have a sum of 1333.

The mean, $\bar{x} = \dfrac{\Sigma xf}{\Sigma f} = \dfrac{1333}{40} = 33.325$.

Combined sets of data

There are many different ways to combine sets of data. However, here we do this by simply considering all of their values together. To find the mean of two combined sets, we divide the sum of all their values by the total number of values in the two sets.

For example, by combining the dataset 1, 2, 3, 4 with the dataset 4, 5, 6, we obtain a new set of data that has seven values in it: 1, 2, 3, 4, 4, 5, 6. Note that the value 4 appears twice. Individually, the sets have means of $\dfrac{1+2+3+4}{4} = 2.5$ and $\dfrac{4+5+6}{3} = 5$. The combined sets have a mean of $\dfrac{1+2+3+4+4+5+6}{4+3} = 3\frac{4}{7}$.

TIP

Note that the mean of the combined sets is not $\dfrac{2.5 + 5}{2}$.

WORKED EXAMPLE 2.5

A large bag of sweets claims to contain 72 sweets, having a total mass of 852.4 g.

A small bag of sweets claims to contain 24 sweets, having a total mass of 282.8 g.

What is the mean mass of all the sweets together?

Answer

Total number of sweets = 72 + 24 = 96.

Total mass of sweets = 852.4 + 282.8 = 1135.2 g

Mean mass = $\dfrac{1135.2}{96}$ = 11.825 g

We find the total number of sweets and their total mass.

TIP

Our answer assumes that the masses given are accurate to 1 decimal place; that the numbers of sweets given are accurate; and that the masses of the bags are not included in the given totals.

WORKED EXAMPLE 2.6

A family has 38 films on DVD with a mean playing time of 1 hour 32 minutes. They also have 26 films on video cassette, with a mean playing time of 2 hours 4 minutes. Find the mean playing time of all the films in their collection.

Answer

$38 + 26 = 64$ films

$(1\,\text{h}\,32\,\text{min} \times 38) + (2\,\text{h}\,4\,\text{min} \times 26) = (92 \times 38) + (124 \times 26)$ .... We find the total number of films and their total playing time.
$\qquad\qquad\qquad\qquad\qquad\qquad\quad = 6720\,\text{min}$

Mean playing time $= \dfrac{6720}{64}$ .... The 64 films have a total playing time of 6720 minutes.

$\qquad\qquad\qquad\quad = 105\,\text{min}$ or $1\,\text{h}\,45\,\text{min}$.

EXPLORE 2.2

In Worked example 2.6, the mean playing time of 105 minutes is not equal to $\dfrac{92 + 124}{2}$.

The mean of A and $B \neq \dfrac{\text{mean of } A + \text{mean of } B}{2}$ but this is not always the case.

Suppose two sets of data, A and B, have m and n values with means $\dfrac{\Sigma A}{m}$ and $\dfrac{\Sigma B}{n}$, respectively.

In what situations will the mean of A and B together be equal to $\dfrac{\text{mean of } A + \text{mean of } B}{2}$?

> **TIP**
>
> The symbol $\neq$ means 'is not equal to'.

Means from grouped frequency tables

When data are presented in a grouped frequency table or illustrated in a histogram or cumulative frequency graph, we lose information about the raw values. For this reason we cannot determine the mean exactly but we can calculate an estimate of the mean. We do this by using mid-values to represent the values in each class.

We use the formula $\bar{x} = \dfrac{\Sigma xf}{\Sigma f}$, given in Key point 2.2, to calculate an estimate of the mean, where x now represents the class mid-values.

WORKED EXAMPLE 2.7

Coconuts are packed into 75 crates, with 40 of a similar size in each crate.

46 crates contain coconuts with a total mass from 20 up to but not including 25 kg.

22 crates contain coconuts with a total mass from 25 up to but not including 40 kg.

7 crates contain coconuts with a total mass from 40 up to but not including 54 kg.

a Calculate an estimate of the mean mass of a crate of coconuts.

b Use your answer to part **a** to estimate the mean mass of a coconut.

Answer

a

Mass (kg)	20–	25–	40–54	
No. crates (f)	46	22	7	$\Sigma f = 75$
Mid-value (x)	22.5	32.5	47.0	
xf	1035	715	329	$\Sigma xf = 2079$

We tabulate the data to include class mid-values, x, and the products xf.

Estimate for mean, $\bar{x} = \dfrac{\Sigma xf}{\Sigma f} = \dfrac{2079}{75} = 27.72$ kg

We estimate that the 75 crates have a total mass of 2079 kg.

b Estimate for the mean mass of a coconut $= \dfrac{27.72}{40}$
$= 0.693$ kg

We divide our answer to part **a** by 40 because each crate contains 40 coconuts.

When gaps appear between classes of grouped data, class boundaries should be used to find class mid-values. The following example shows a situation in which it is common for incorrect boundaries and mid-values to be used.

WORKED EXAMPLE 2.8

Calculate an estimate of the mean age of a group of 50 students, where there are sixteen 18-year-olds, twenty 19-year-olds and fourteen who are either 20 or 21 years old.

Answer

Age (A years)	Mid-value (x)	No. students (f)	xf
$18 \leq A < 19$	18.5	16	296
$19 \leq A < 20$	19.5	20	390
$20 \leq A < 22$	21.0	14	294
		$\Sigma f = 50$	$\Sigma xf = 980$

The 18-year-olds are all aged from 18 up to but not including 19 years.

The 19-year-olds are all aged from 19 up to but not including 20 years.

The 20- and 21-year-olds are all aged from 20 up to but not including 22 years.

The age groups and necessary totals are shown in the table.

Estimate of the mean age is $\dfrac{\Sigma xf}{\Sigma f} = \dfrac{980}{50}$
$= 19.6$ years

Note: boundary values for the true mean are:

lower $= \dfrac{18 \times 16 + 19 \times 20 + 20 \times 14}{50} = 18.96$

upper $= \dfrac{19 \times 16 + 20 \times 20 + 22 \times 14}{50} = 20.24$

TIP

Incorrect mid-values of 18, 19 and 20.5 give an incorrect estimated mean of 19.1 years.

FAST FORWARD

We will see how the mean is used to calculate the variance and standard deviation of a set of data in Chapter 3, Section 3.3.

EXERCISE 2B

1. Calculate the mean of the following sets of numbers:

   a 28, 16, 83, 72, 105, 55, 6 and 35

   b 7.3, 8.6, 11.7, 9.1, 1.7 and 4.2

   c $3\frac{1}{2}, 5\frac{1}{4}, 9\frac{3}{4}, -4\frac{1}{4}$ and $7\frac{3}{8}$.

2. a The mean of 15, 31, 47, 83, 97, 119 and p^2 is 63. Find the possible values of p.

   b The mean of 6, 29, 3, 14, q, $(q+8)$, q^2 and $(10-q)$ is 20. Find the possible values of q.

3. Given that:

   a $n = 14$ and $\Sigma x = 325.5$, find $\bar{x}$.

   b $n = 45$ and $\bar{y} = 23.6$, find the value of Σy.

   c $\Sigma z = 4598$ and $\bar{z} = 52.25$, find the number of values in the set of data.

   d $\Sigma xf = 86$ and $\bar{x} = 7\frac{1}{6}$, find the value of Σf.

   e $\Sigma f = 135$ and $\bar{x} = 0.842$, find the value of Σxf.

4. Find the mean of x and of y given in the following tables.

   a

x	18.0	18.5	19.0	19.5	20.0
f	8	10	17	24	1

   b

y	3.62	3.65	3.68	3.71	3.74
f	127	209	322	291	251

5. For the data given in the following table, it is given that $\bar{q} = 8\frac{5}{9}$.

q	7	8	9	10
f	9	13	a	11

   Calculate the value of a.

6. Calculate an estimate of the mean of x and of y given in the following tables.

   a

x	$0 \leq x < 2$	$2 \leq x < 4$	$4 \leq x < 8$	$8 \leq x < 14$
f	8	9	11	2

   b

y	$13 \leq y < 16$	$16 \leq y < 21$	$21 \leq y < 28$	$28 \leq y < 33$	$33 \leq y < 36$
f	7	17	29	16	11

7. An examination was taken by 50 students. The 22 boys scored a mean of 71% and the girls scored a mean of 76%. Find the mean score of all the students.

8. A company employs 12 drivers. Their mean monthly salary is $1950. A new driver is employed and the mean monthly salary falls by $8. Find the monthly salary of the new driver.

9. The mean age of the 16 members of a karate club is 26 years and 3 months. One member leaves the club and the mean age of those remaining is 26 years. Find the age of the member who left the club. Give a reason why your answer might not be very accurate.

Chapter 2: Measures of central tendency

M 10 The following table shows the hourly rates of pay, in dollars, of a company's employees.

Hourly rate ($)	6	7	8	109
No. employees (f)	8	11	17	1

   a Is the mean a good average to use here? Give a reason for your answer.

   b Find the mean rate of pay for the majority of the employees.

PS P 11 A train makes a non-stop journey from one city to another and back again each day. Over a period of 30 days, the mean number of passengers per journey is exactly 61.5. Exact one-way ticket prices paid by these passengers are given by percentage in the following table.

Price ($)	34	38	45
Passengers (%)	30	41	29

   a Calculate the total revenue from ticket sales, and explain why your answer is an approximation.

   b The minimum and maximum possible revenues differ by $$k$. Find the value of k.

12 The heights, in centimetres, of 54 children are represented in the following diagram.

The children are split into two equal-sized groups: a 'tall half' and a 'short half'.

Calculate an estimate of the difference between the mean heights of these two groups of children.

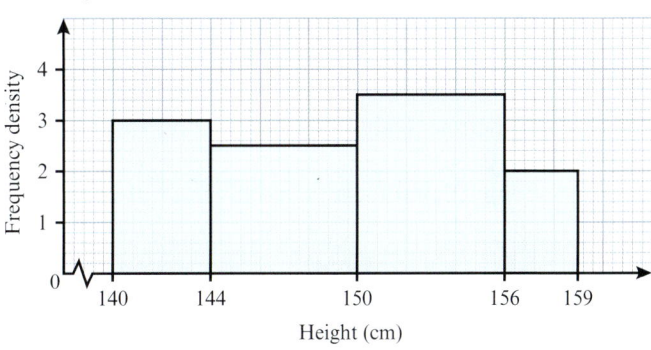

M 13 The following table summarises the number of tomatoes produced by the plants in the plots on a farm.

No. tomatoes	20–29	30–49	50–79	80–100
No. plots (f)	329	413	704	258

   a Calculate an estimate of the mean number of tomatoes produced by these plots.

   b The tomatoes are weighed accurately and their mean mass is found to be 156.50 grams. At market they are sold for $3.20 per kilogram and the total revenue is $50 350. Find the actual mean number of tomatoes produced per plot.

   c Why could your answer to part **b** be inaccurate?

14 Twenty boys and girls were each asked how many aunts and uncles they have. The entry 4/5 in the following table, for example, shows that 4 boys and 5 girls each have 3 aunts and 2 uncles.

		Aunts			
B/G		0	1	2	3
Uncles	0	1/0	0/2	2/1	1/1
	1	0/0	3/4	0/4	0/0
	2	0/0	1/0	7/11	4/5
	3	1/0	0/0	0/1	0/1

   a Find the mean number of uncles that the boys have.

   b For the boys and girls together, calculate the mean number of:

      i aunts   ii aunts and uncles.

   c Suggest an alternative way of presenting the data so that the calculations in parts **a** and **b** would be simpler to make.

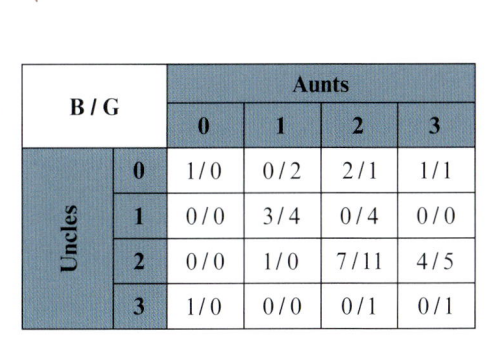

15 **PS** A calculated estimate of the mean capacity of 120 refrigerators stored at a warehouse is 348 litres. The capacities are given in the following table.

Capacity (litres)	160–	200–	320–	400– p
No. refrigerators (f)	12	28	48	32

A delivery of n new refrigerators, all with capacities between 200 and 320 litres arrives at the warehouse. This causes the mean capacity to decrease by 8 litres. Find the value of n and state what assumptions you are making in your calculations.

16 **PS** A carpet fitter is employed to fit carpet in each of the 72 guest bedrooms at a new hotel. The following table shows how many rooms were completed during the first 10 days of work.

No. rooms completed	5	6 or 7
No. days (f)	2	8

Based on these figures, estimate how many more days it will take to finish the job. What assumptions are you making in your calculations?

17 **PS** In the figure opposite, a square of side 8 cm is joined edge-to-edge to a semicircle, with centre O. P is 2 cm from O on the figure's axis of symmetry.

Points X and Y are fixed but the position of Z is variable on the shape's perimeter.

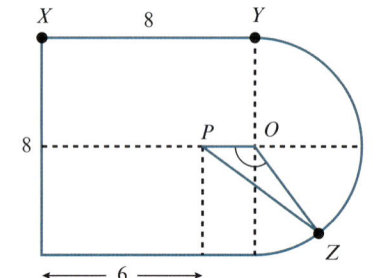

a Find the mean distance from P to X, Y and Z when angle POZ is equal to:

   i 180°   ii 135°.

b Find obtuse angle POZ, so that the mean distance from P to X, Y and Z is identical to the mean distance from P to X and Y.

EXPLORE 2.3

Six cards, numbered 1, 2, 3, 4, 5 and 6, are placed in a bag, as shown.

15 different pairs of cards can be selected without replacement from the bag. Three of these pairs are {2, 3}, {6, 4} and {5, 1}.

Make a list of all 15 unordered pairs and find the mean of each. We will denote these mean values by X_2.

Choose a suitable method to represent the values of X_2 and their frequencies. Find $\bar{X}_2$, the mean of the values of X_2.

Repeat the process described above for each of the following:

- the six possible selections of five cards, denoting their means by X_5
- the 15 possible selections of four cards, denoting their means by X_4
- the 20 possible selections of three cards, denoting their means by X_3
- the six possible selections of one card, denoting their means by X_1.

FAST FORWARD

We will study the number of ways of selecting objects in Chapter 5, Section 5.3.

What does the single value of X_6 represent?

Do the values of X_1, X_2, X_3, X_4 and X_5 have anything in common?

Can you suggest reasons for any of the common features that you observe?

Investigate the values of X_r when there are a different number of consecutively numbered cards in the bag.

Coded data

To code a set of data, we can *transform* all of its values by addition of a positive or negative constant. The result of doing this produces a set of **coded** data.

One reason for coding is to make the numbers easier to handle when performing manual calculations. Also, it is sometimes easier to work with coded data than with the original data (by arranging the mean to be a convenient number, such as zero, for example).

To find the mean of 101, 103, 104, 109 and 113, for example, we can use the values 1, 3, 4, 9 and 13.

Our x values are 101, 103, 104, 109 and 113, so 1, 3, 4, 9 and 13 are corresponding values of $(x - 100)$.

Mean of the coded values is $\dfrac{\Sigma(x-100)}{5} = \dfrac{1+3+4+9+13}{5} = 6$.

We subtracted 100 from each x value, so we simply add 100 to the mean of the coded values to find the mean of x.

Mean(x) = mean$(x - 100) + 100 = 106$ or $\bar{x} = \dfrac{\Sigma(x-100)}{5} + 100 = 106$

Refer to the following diagram. If we add $-b$ to the set of x values, they are all translated by $-b$ and so is their mean.

So, mean$(x - b) = \bar{x} - b$.

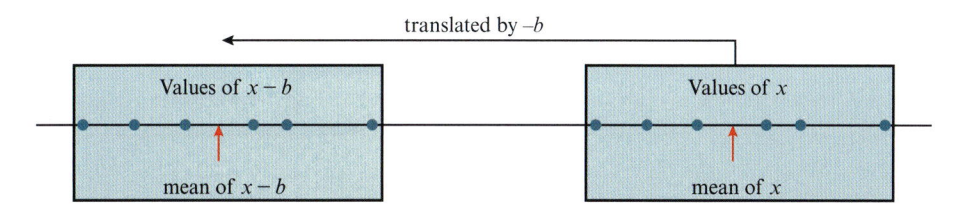

translated by $-b$

Values of $x - b$ Values of x

mean of $x - b$ mean of x

> **FAST FORWARD**
>
> We will study the standard normal variable, which has a mean of 0, in Chapter 8, Section 8.2.

> **FAST FORWARD**
>
> We will see how to use coded totals, such as $\Sigma(x-b)$ and $\Sigma(x-b)^2$, to find measures of variation in Chapter 3, Section 3.3.

> **TIP**
>
> If we remove the bracket from $\Sigma(x-b)$, we obtain $\Sigma x - \Sigma b$. The term Σb means 'the sum of all the bs' and there are n of them, so $\Sigma(x-b) = \Sigma x - nb$.

KEY POINT 2.3

For ungrouped data, $\bar{x} = \dfrac{\Sigma(x-b)}{n} + b$.

For grouped data, $\bar{x} = \dfrac{\Sigma(x-b)f}{\Sigma f} + b$.

These formulae can be summarised by writing $\bar{x} = $ mean$(x-b) + b$.

For two datasets coded as $(x - a)$ and $(y - b)$, we can use the totals Σx and Σy to find the mean of the combined set of values of x and y.

WORKED EXAMPLE 2.9

The exact age of an individual boy is denoted by b, and the exact age of an individual girl is denoted by g.

Exactly 5 years ago, the sum of the ages of 10 boys was 127.0 years, so $\Sigma(b - 5) = 127.0$.

In exactly 5 years' time, the sum of the ages of 15 girls will be 351.0 years, so $\Sigma(g + 5) = 351.0$.

Find the mean age today of

    a the 10 boys   **b** the 15 girls   **c** the 10 boys and 15 girls combined.

Answers

a $\bar{b} = \dfrac{127}{10} + 5 = 17.7$ years   We update the boys' past mean age by addition.

$\Sigma(b - 5) = \Sigma b - (10 \times 5) = 127$, so   Alternatively, we expand the brackets.

$\Sigma b = 127 + 50 = 177$ and $\bar{b} = \dfrac{177}{10} = 17.7$ years

b $\bar{g} = \dfrac{351}{15} - 5 = 18.4$ years   We backdate the girls' future mean age by subtraction.

$\Sigma(g + 5) = \Sigma g + (15 \times 5) = 351$, so   Alternatively, we expand the brackets.

$\Sigma g = 351 - 75 = 276$ and $\bar{g} = \dfrac{276}{15} = 18.4$ years

c $\dfrac{\Sigma b + \Sigma g}{10 + 15} = \dfrac{177 + 276}{25} = 18.12$ years

WORKED EXAMPLE 2.10

Forty values of x are coded in the following table.

$x - 3$	0–	18–	24–32
Frequency	9	13	18

Calculate an estimate of the mean value of x.

Answer

$\bar{x} = \dfrac{\Sigma(x - 3)f}{\Sigma f} + 3$

$= \dfrac{(9 \times 9) + (21 \times 13) + (28 \times 18)}{40} + 3$

$= 24.45$

We calculate an estimate for the mean of the coded data using class mid-values of 9, 21 and 28, and then add 3 to obtain our estimate for $\bar{x}$.

TIP

It is not necessary to *decode* the values of $x - 3$.

Chapter 2: Measures of central tendency

EXERCISE 2C

1. For 10 values denoted x, it is given that $\bar{x} = 7.4$. Find:

   a Σx b $\Sigma(x+2)$ c $\Sigma(x-1)$

2. Twenty-five values of z are such that $\Sigma(z-7) = 275$. Find $\bar{z}$.

3. Given $\bar{q} = 22$ and $\Sigma(q-4) = 3672$, find the number of values of q.

4. The lengths of 2500 bolts, x mm, are summarised by $\Sigma(x-40) = 875$. Find the mean length of the bolts.

5. Six data values are coded by subtracting 13 from each. Five of the coded values are 9.3, 5.4, 3.9, 7.6 and 2.2, and the mean of the six data values is 17.6.

   Find the sixth coded value.

6. The SD card slots on digital cameras are designed to accommodate a card of up to 24 mm in width. Due to low sales figures, a manufacturer suspects that the machine used to cut the cards needs to be recalibrated. The widths, w mm, of 400 of these cards were measured and are coded in the following table, where $x = w - 24$.

$w - 24$ (mm)	$-0.15 \leqslant x < -0.1$	$-0.1 \leqslant x < 0$	$0 \leqslant x < 0.1$	$0.1 \leqslant x < 0.2$
No. cards (f)	32	360	6	2

   a Suggest a reason why the widths have been coded in this way.

   b What percentage of the SD cards are too wide to fit into the slots?

   c Use the coded data to estimate the mean width of these 400 cards.

7. Sixteen bank accounts have been accidentally under-credited by the following amounts, denoted by $\$x$.

   917.95   917.98   918.03   917.97   918.01   917.94   918.05   918.07

   918.02   917.93   918.01   917.88   918.10   917.85   918.11   917.94

   To calculate $\bar{x}$ manually, Fidel and Ramon code these figures using $(x - 917)$ and $(x - 920)$, respectively.

   Who has the simpler maths to do? Explain your answer.

8. Throughout her career, an athlete has been timed in 120 of her 400-metre races. Her times, denoted by t seconds, were recorded on indoor tracks 45 times and are summarised by $\Sigma(t - 60) = 83.7$, and on outdoor tracks where $\Sigma(t - 65) = -38.7$. Calculate her average 400-metre running time and comment on the accuracy of your answer.

9. All the interior angles of n triangular metal plates, denoted by $y°$, are measured.

   a State the number of angles measured and write down the value of $\bar{y}$.

   b Hence, or otherwise, find the value of $\Sigma(y - 30)$.

10. A dataset of 20 values is denoted by x where $\Sigma(x-1) = 58$. Another dataset of 30 values is denoted by y where $\Sigma(y-2) = 36$. Find the mean of the 50 values of x and y.

> **FAST FORWARD**
>
> You will study the mean of linear combinations of random variables in the Probability & Statistics 2 Coursebook, Chapter 3.

11 Students investigated the prices in dollars ($) of 1 litre bottles of a certain drink at 24 shops in a town and at 16 shops in surrounding villages. Denoting the town prices by t and the village prices by v, the students' data are summarised by the totals $\Sigma(t-1.1) = 1.44$ and $\Sigma(v-1.2) = 0.56$.

Find the mean price of 1 litre of this drink at all the shops at which the students collected their data.

E A set of data can be coded by multiplication as well as by addition of a constant.

Suppose the monthly take-home salaries of four teachers are $3600, $4200, $3700 and $4500, which have mean $\bar{x} = \$4000$.

What happens to the mean if all the teachers receive a 10% increase but must pay an extra $50 in tax each month?

To find their new take-home salaries, we multiply the current salaries by 1.1 and then subtract 50.

The new take-home salaries are $3910, $4570, $4020 and $4900.

The mean is $\dfrac{3910 + 4570 + 4020 + 4900}{4} = \$4350$.

The original data, x, has been coded by multiplication and by addition as $1.1x - 50$.

The mean of the coded data is 4350, which is equal to $(1.1 \times 4000) - 50$, where $4000 = \bar{x}$.

Data coded as $ax - b$ has a mean of $a\bar{x} - b$.

To find $\bar{x}$ from a total such as $\Sigma(ax - b)$, we can find the mean of the coded data, then undo '$-b$' and undo '$\times a$', in that order. That is:

$\bar{x} = (4350 + 50) \div 1.1$ or $\dfrac{1}{1.1} \times (4350 + 50) = 4000$.

> **KEY POINT 2.4**
>
> For ungrouped data,
> $$\bar{x} = \dfrac{1}{a}\left[\dfrac{\Sigma(ax-b)}{n} + b\right].$$
>
> For grouped data,
> $$\bar{x} = \dfrac{1}{a}\left[\dfrac{\Sigma(ax-b)f}{\Sigma f} + b\right].$$
>
> These formulae can be summarised by writing
> $$\bar{x} = \dfrac{1}{a} \times [\text{mean}(ax-b) + b].$$

> **TIP**
>
> $\Sigma(ax - b)$ can be rewritten as $a\Sigma x - nb$.

WORKED EXAMPLE 2.11

The total area of cloth produced at a textile factory is denoted by Σx and is measured in square metres. Find an expression in x for the area of cloth produced in square centimetres.

Answer

$1\,\text{m} = 100\,\text{cm}$

$1\,\text{m}^2 = 100^2\,\text{cm}^2 = 10\,000\,\text{cm}^2$ We convert the measurements of x from m² to cm².

Total area, in square centimetres, is $\Sigma 10\,000x$ or $10\,000\Sigma x$.

WORKED EXAMPLE 2.12

For the 20 values of x summarised by $\Sigma(2x - 3) = 104$, find $\bar{x}$.

Answer

$\dfrac{104}{20} = 5.2$ We first find the mean of the coded values.

$\bar{x} = \dfrac{5.2 + 3}{2} = 4.1$ Knowing that $2\bar{x} - 3 = 5.2$, we undo the '-3' and then undo the '$\times 2$', in that order, to find $\bar{x}$.

Chapter 2: Measures of central tendency

$\Sigma(2x-3) = 104$
$2\Sigma x - (20 \times 3) = 104$
$\Sigma x = 82$

$\bar{x} = \dfrac{82}{20} = 4.1$

Alternatively, we can expand the brackets in $\Sigma(2x-3)$, which allows us to find the value of Σx.

FAST FORWARD

We will see how to use coded totals such as $\Sigma(ax-b)$ and $\Sigma(ax-b)^2$ to find measures of variation in Chapter 3, Section 3.3.

EXERCISE 2D

1 The masses, x kg, of 12 objects are such that $\bar{x} = 0.475$. Find the value of $\Sigma 1000x$ and state what it represents.

2 The total mass of gold extracted from a mine is denoted by Σx, which is measured in grams. Find an expression in x for the total mass in:

   a carats, given that 1 carat is equivalent to 200 milligrams

   b kilograms.

3 The area of land used for growing wheat in a region is denoted by Σw hectares. Find an expression in w for the total area in square kilometres, given that 1 hectare is equivalent to $10\,000\,\text{m}^2$.

4 Speeds, measured in metres per second, are denoted by x. Find the constant k such that kx denotes the speeds in kilometres per hour.

5 The wind speeds, x miles per hour (mph), were measured at a coastal location at midday on 40 consecutive days and are presented in the following table.

Speed (x mph)	$15 \leqslant x < 17$	$17 \leqslant x < 20$	$20 \leqslant x < 24$	$24 \leqslant x < 25$
No. days (f)	9	13	14	4

Abel wishes to calculate an estimate of the mean wind speed in kilometres per hour (km/h). He knows that a distance of 5 miles is approximately equal to 8 km.

   a Explain how Abel can calculate his estimate without converting the given boundary values from miles per hour to kilometres per hour.

   b Use the wind speeds in mph to estimate the mean wind speed in km/h.

6 Given that 15 values of x are such that $\Sigma(3x-2) = 528$, find $\bar{x}$ and find the value of b such that $\Sigma(0.5x - b) = 138$.

7 For 20 values of y, it is given that $\Sigma(ax - b) = 400$ and $\Sigma(bx - a) = 545$. Given also that $\bar{x} = 6.25$, find the value of a and of b.

P 8 The midpoint of the line segment between A and B is at $(5.2, -1.2)$.

Find the coordinates of the midpoint after the following transformations have been applied to A and to B.

   a T: Translation by the vector $\begin{pmatrix} -7 \\ 4 \end{pmatrix}$.

b E: Enlargement through the origin with scale factor 5.

c Transformations T and E are carried out one after the other. Investigate whether the location of the mid-point of AB is independent of the order in which the transformations are carried out.

M 9 Five investors are repaid, each with their initial investment increased by $p\%$ plus a fixed 'thank you' bonus of $\$q$. The woman who invested $\$20\,000$ is repaid double her investment and the man who invested $\$7500$ is repaid triple his investment. Find the total amount that the five people invested, given that the mean amount repaid to them was $\$33\,000$.

Do you think the method of repayment is fair? Give a reason for your answer.

PS 10 One of the units used to measure pressure is pounds per square inch (psi). The mean pressure in the four tyres of a particular vehicle is denoted by $\bar{x}$ psi. Given that 1 pound is approximately equal to 0.4536 kg and that 1 metre is approximately equal to 39.37 inches, express the sum of the pressures in the four tyres of this vehicle in grams per cm².

2.3 The median

You will recall that the median splits a set of data into two parts with an equal number of values in each part: a bottom half and a top half. In a set of n ordered values, the median is at the value half-way between the 1st and the nth.

Consider a DIY store that opens for 12 hours on Monday and for 15 hours on Saturday. The numbers of customers served during each hour on Monday and on Saturday last week are shown in the following back-to-back stem-and-leaf diagram.

```
Monday (12) |   | Saturday (15)   Key: 0 | 2 | 2
8 6 3 1 0 0 | 2 | 2 3 4 6         represents 20 customers on
    4 3 1 1 0 | 3 | 5 5 6 8 9 9   Monday and 22 customers
            1 | 4 | 0 1 3 7 9     on Saturday
```

To find the median number of customers served on each of these days, we need to find their positions in the ordered rows of the back-to-back stem-and-leaf diagram.

For Saturday, there are $n = 15$ values arranged in ascending order from top to bottom and from left to right. The median is at the $\left(\dfrac{n+1}{2}\right)$th $= \left(\dfrac{15+1}{2}\right) = 8$th value.

In the first row, we have the 1st to 4th values, and in the second row we have the 5th to 10th values, so the 8th value is 38.

The median number of customers on Saturday was 38.

For Monday, there are $n = 12$ values arranged in ascending order from top to bottom and from right to left. The median is at the $\left(\dfrac{n+1}{2}\right)$th $= \left(\dfrac{12+1}{2}\right)$th $= 6.5$th value, so we locate the median mid-way between the 6th and 7th values.

In the first row, we have the 1st to 6th values and the 6th is 28.

The first value in the second row is the 7th value, which is 30.

The median number of customers on Monday was $\dfrac{28+30}{2} = 29$.

When data appear in an ordered frequency table of individual values, we can use cumulative frequencies to investigate the positions of the values, knowing that the median is at the $\left(\dfrac{n+1}{2}\right)$th value.

KEY POINT 2.5

For n ordered values, the median is at the $\left(\dfrac{n+1}{2}\right)$th value.

For even values of n, the median is the mean of the two middle values.

TIP

We can find the 8th value by counting down and left to right from 22 or by counting up and right to left from 49.

TIP

Take care when locating values at the left side of a back-to-back stem-and-leaf diagram; they ascend from right to left, and descend from left to right, as we move along each row.

TIP

n is equal to the total frequency Σf.

WORKED EXAMPLE 2.13

The following table shows 65 ungrouped readings of x. Cumulative frequencies and the positions of the readings are also shown. Find the median value of x.

x	f	cf	Positions
40	11	11	1st to 11th
41	23	34	12th to 34th
42	19	53	35th to 53rd
43	8	61	54th to 61st
44	4	65	62nd to 65th

Answer

Median value of x is 41.

The total frequency is 65, and $\frac{n+1}{2} = \frac{65+1}{2} = 33$, so the median is at the 33rd value. From the table, we see that the 12th to 34th values are all equal to 41.

> **TIP**
>
> Frequencies must be taken in account here. Although 41 is not the middle of the five values of x, it is the middle of the 65 readings.

Estimating the median

In large datasets and in sets of continuous data, values are grouped and the actual values cannot be seen. This means that we cannot find the exact value of the median but we can estimate it. The method we use to estimate the median for this type of data is by reading its value from a cumulative frequency graph. We estimate the median to be the value whose cumulative frequency is equal to half of the total frequency.

Consider the masses of 300 museum artefacts, which are represented in the following cumulative frequency graph.

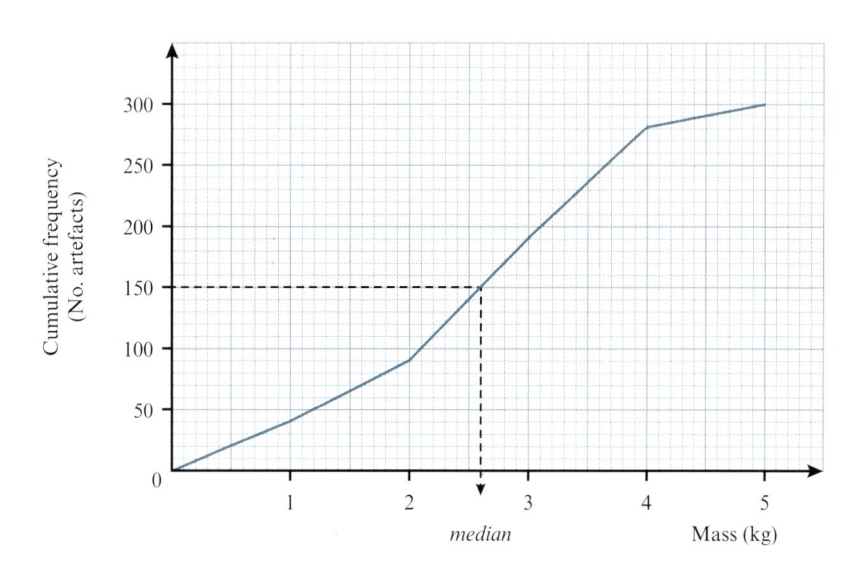

> **REWIND**
>
> We studied cumulative frequency graphs in Chapter 1, Section 1.4.

> **KEY POINT 2.6**
>
> On a cumulative frequency graph with total frequency $n = \Sigma f$, the median is at the $\frac{n}{2}$ th value.

> **TIP**
>
> The graph is only an estimate, so we use $\frac{n}{2}$ rather than $\frac{n+1}{2}$ to estimate the median. This ensures that we arrive at the same position for the median whether we count up from the bottom or down from the top of the cumulative frequency axis.

The set of data has a total of $n = 300$ values.

An estimate for the median is the mass of the $\frac{n}{2} = \frac{300}{2} = 150$th artefact.

Cambridge International AS & A Level Mathematics: Probability & Statistics 1

We draw a horizontal line from a cumulative frequency value of 150 to the graph. Then, at the point of intersection, we draw a perpendicular, vertical line down to the axis showing the masses.

Reading from the graph, we see that the median mass is approximately 2.6 kg.

> **TIP**
>
> Do not confuse the median's position (150th) with its value (2.6 kg).

DID YOU KNOW?

The concept of representing many different measurements with one representative value is quite a recent invention. There are no historical examples of the mean, median or mode being used before the 17th century.

In trying to find the longitude of Ghanza in modern-day Afghanistan, and in studying the characteristics of metals, the 11th century Persian Al-Biruni is one of the earliest known users of a method for finding a representative measure. He used the number in the middle of the smallest and largest values (what we would call the *mid-range*) ignoring all but the minimum and maximum values.

The mid-range was used by Isaac Newton and also by explorers in the 17th and 18th centuries to estimate their geographic positions. It is likely that measuring magnetic declination (i.e. the variation in the angle of magnetic north from true north) played a large part in the growth of the mean's popularity.

> **FAST FORWARD**
>
> We will use cumulative frequency graphs to estimate the quartiles, the interquartile range and percentiles in Chapter 3, Section 3.2.

> **FAST FORWARD**
>
> The mid-range, as you will discover in Chapter 3, Section 3.2, is not the same as the median.

Choosing an appropriate average

Selecting the most appropriate average to represent the values in a set of data is a matter for discussion in most situations. Just as it may be possible to choose an average that represents the data well, so it is often possible to choose an average that badly misrepresents the data. The purpose and motives behind choosing an average value must also be considered as part of the equation.

Consider a student whose marks out of 20 in 10 tests are: 3, 4, 6, 7, 8, 11, 12, 13, 17 and 17.

The three averages for this set of data are: mode = 17, mean = 9.8 and median = 9.5.

If the student wishes to impress their friends (or parents), they are most likely to use the mode as the average because it is the highest of the three. Using either the mean or median would suggest that, on average, the student scored fewer than half marks on these tests.

Some of the features of the measures of central tendency are given in the following table.

	Advantages	Disadvantages
Mode	Unlikely to be affected by **extreme values**. Useful to manufacturers that need to know the most popular styles and sizes. Can be used for all sets of qualitative data.	Ignores most values. Rarely used in further calculations.
Mean	Takes all values into account. Frequently used in further calculations. The most commonly understood average. Can be used to find the sum of the data values.	Cannot be found unless all values are known. Likely to be affected by extreme values.
Median	Can be found without knowing all of the values. Relatively unaffected by extreme values.	Only takes account of the order of the values and so ignores most of them.

As an example of the effect of an extreme value, consider the dataset 40, 40, 70, 100, 130 and 250. If we increase the largest value from 250 to 880, the mode and median are unchanged (i.e. 40 and 85), but the mean increases by 100% from 105 to 210.

Although the median is usually unaffected by extreme values, this is not always the case, as the *Libor scandal* shows.

> **DID YOU KNOW?**
>
> LIBOR (London Interbank Offered Rates) are average interest rates that the world's leading banks charge each other for short-term loans. They determine the prices that people and businesses around the world pay for loans or receive for their savings. They underpin over US $450 trillion worth of investments and are used to assess the health of the world's financial system.
>
> A   B   C   D
> 2.6% 2.8% 3.0% 3.1%
> LIBOR = 2.9%
>
> The highest and lowest 25% of the daily rates submitted by a small group of leading banks are discarded and a LIBOR is then fixed as the mean of the middle 50%. The above diagram shows a simple example.
>
> Consider how the LIBOR would be affected if bank D submitted a rate of 2.5% instead of 3.1%.
>
> Several leading banks have been found guilty of manipulating the LIBOR by submitting false rates, which has so far resulted in them being fined over US $9 billion.
>
> You can find out more about the LIBOR scandal by searching news websites.

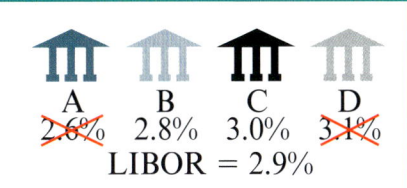

Consider the number of days taken by a courier company to deliver 100 packages, as given in the following table and represented in the bar chart.

No. days	1	2	3	4	5	6
No. packages (f)	10	40	25	14	8	3

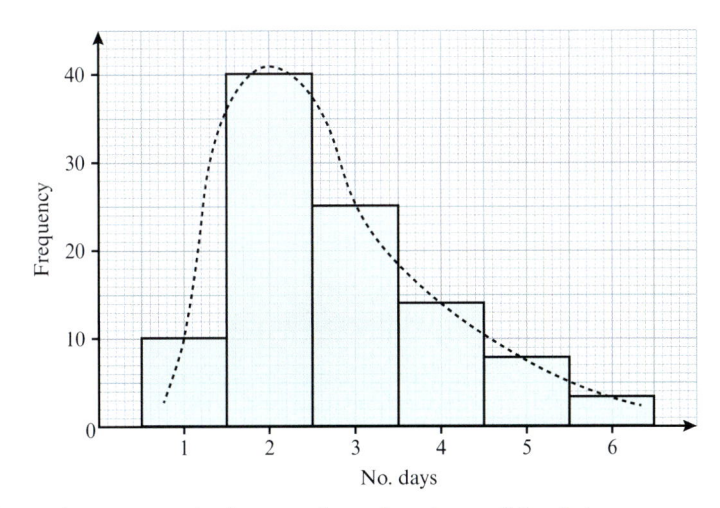

A curve has been drawn over the bars to show the *shape* of the data.

The mode is 2 days.

The median is between the 50th and 51st values, which is 2.5 days.

The mean $= \dfrac{(1\times 10)+(2\times 40)+(3\times 25)+(4\times 14)+(5\times 8)+(6\times 3)}{100} = 2.79$ days

Cambridge International AS & A Level Mathematics: Probability & Statistics 1

The mean is the largest average and is to the side of the curve's longer tail.

The mode is the smallest average and is to the side of the curve's shorter tail.

The median is between the mode and the mean.

A set of data that is not symmetrical is said to be **skewed**. When the curve's longer tail is to the side of the larger values, as in the previous bar chart, the data are said to be positively skewed. When the longer tail is to the side of the smaller values, the data are said to be negatively skewed.

Generally, we find that:

Mode < median < mean when the data are positively skewed.

Mean < median < mode when the data are negatively skewed.

> **FAST FORWARD**
>
> In Chapter 3, we will use a measure of central tendency and a measure of variation to better describe the values in a set of data.
>
> In Chapter 8, we will study sets of data called *normal distributions* in which the mode, mean and median are equal.

EXERCISE 2E

1 The number of patients treated each day by a dentist during a 20-day period is shown in the following stem-and-leaf diagram.

```
0 | 4 4 4 5 6 6 6 7         Key: 1 | 5
1 | 4 5 5 6 6 7 7 8 8 9     represents
2 | 0 1                     15 patients
```

   a Find the median number of patients.

   b On eight of these 20 days, the dentist arrived late to collect their son from school. If they decide to use their average number of patients as a reason for arriving late, would they use the median or the mean? Explain your answer.

   c Describe a situation in which it would be to the dentist's advantage to use a mode as the average.

2 a Find the median for the values of t given in the following table.

t	7	8	9	10	11	12	13
f	4	7	9	14	16	41	9

   b What feature of the data suggests that $\bar{t}$ is less than the median? Confirm whether or not this is the case.

3 a Find the median and the mode for the values of x given in the following table.

x	4	5	6	7	8
f	14	13	4	12	15

   b Give one positive and one negative aspect of using each of the median and the mode as the average value for x.

   c Some values in the table have been incorrectly recorded as 8 instead of 4. Find the number of incorrectly recorded values, given that the true median of x is 5.5.

4 The following graph illustrates the times taken by 112 people to complete a puzzle.

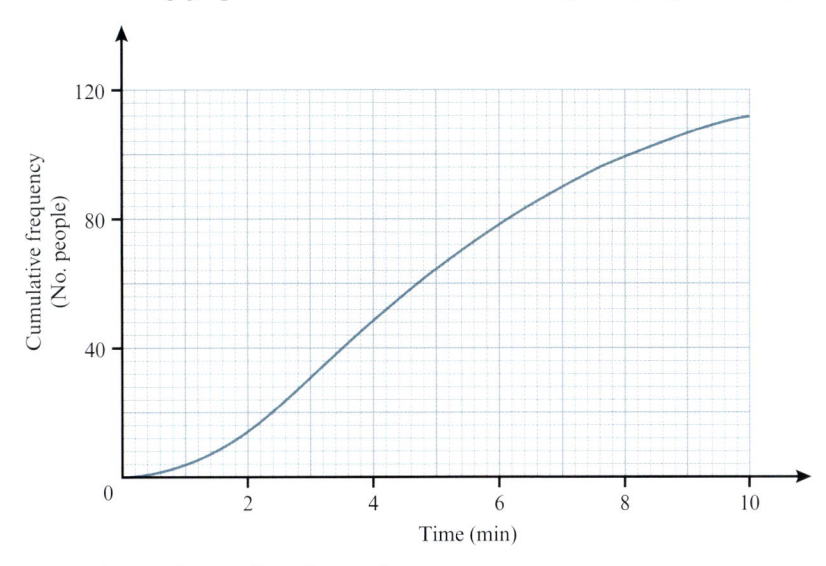

   a Estimate the median time taken.

   b The median is used to divide these people into two groups. Find the median time taken by each of the groups.

5 The masses, m kilograms, of 148 objects are summarised in the following table.

Mass (m kg)	$m < 0$	$m < 0.2$	$m < 0.3$	$m < 0.5$	$m < 0.7$	$m < 0.8$
cf	0	16	28	120	144	148

Construct a cumulative frequency polygon on graph paper, and use it to estimate the number of objects with masses that are:

   a within 0.1 kg of the median

   b more than 200 g from the median.

6 A teacher recorded the quiz marks of eight students as 11, 13, 15, 15, 17, 18, 19 and 20.

   They later realised that there was a typing error, so they changed the mark of 11 to 1.

   Investigate what effect this change has on the mode, mean and median of the students' marks.

7 The following table shows the lifetimes, to the nearest 10 days, of a certain brand of light bulb.

Lifetime (days)	90–100	110–120	130–140	150–160	170–190	200–220	230–260
No. light bulbs (f)	12	28	54	63	41	16	6

   a Use upper class boundaries to represent the data in a cumulative frequency graph and estimate the median lifetime of the light bulbs.

   b How might the manufacturer choose a value to use as the average lifetime of the light bulbs in a publicity campaign? Based on the figures in the table, investigate whether it would be to the manufacturer's advantage to use the median or the mean.

8 It is claimed on the packaging of a brand of battery that they can run a standard kitchen clock continuously for 'at least 150 days on average'. Tests are carried out to find the length of time, t hours, that a standard kitchen clock runs using one of these batteries. The results are shown in the following table.

Time (t hours)	$3000 \leq t < 3096$	$3096 \leq t < 3576$	$3576 \leq t < 3768$	$3768 \leq t < 3840$
No. batteries (f)	34	66	117	33

What could the words on the packaging mean? Test the claim by finding the mean, the median and the modal class. What conclusions, if any, can you make about the claim?

9 Homes in a certain neighbourhood have recently sold for $220 000, $242 000, $236 000 and $3 500 000. A potential buyer wants to know the average selling price in the neighbourhood. Which of the mean, median or mode would be more helpful? Explain your answer.

10 A study was carried out on 60 electronic items to find the currents, x amperes, that could be safely passed through them at a fixed voltage before they overheat. The results are given in the two tables below.

Current (x amperes)	0.5	1.5	2.0	3.5	5.0
No. items that do not overheat	60	48	20	6	0

Current (x amperes)	0.5	1.5	2.0	3.5	5.0
No. items that overheat	0	p	q	r	60

a Find the value of p, of q and of r.

b Cumulative frequency graphs are drawn to illustrate the data in both tables.

   i Describe the transformation that maps one graph onto the other.

   ii Explain the significance of the point where the two graphs intersect.

11 The lengths of extra-time, t minutes, played in the first and second halves of 100 football matches are summarised in the following table.

Extra-time (t min)	$t \leq 1$	$t \leq 2$	$t \leq 4$	$t \leq 5$	$t \leq 7$	$t \leq 9$
First halves (cf)	24	62	80	92	97	100
Second halves (cf)	6	17	35	82	93	100

a Explain how you know that the median extra-time played in the second halves is greater than in the first halves.

b The first-half median is exactly 100 seconds.

   i Find the upper boundary value of k, given that the second-half median is k times longer than the first-half median.

   ii Explain why the mean must be greater than the median for the extra-time played in the first halves.

12 Eighty candidates took an examination in Astronomy, for which no candidate scored more than 80%. The examiners suggest that five grades, A, B, C, D and E, should be awarded to these candidates, using upper grade boundaries 64, 50, 36 and 26 for grades B, C, D and E, respectively. In this case, grades A, B, C, D and E, will be awarded in the ratio $1:3:5:4:3$.

a Using the examiners' suggestion, represent the scores in a cumulative frequency polygon and use it to estimate the median score.

b All of the grade boundaries are later reduced by 10%. Estimate how many candidates will be awarded a higher grade because of this.

13 The values of x shown in the following table are to be represented in a bar chart.

x	5	6	7	8	9	10	11
Frequency	2	5	9	10	9	5	2

   a i Sketch a curve that shows the shape of the data.

      ii Find the mode, mean and the median of x.

   b The two smallest values of x (i.e. 5 and 5) are changed to 21 and 31. Investigate the effect that this has on the mode, the mean, the median and on the shape of the curve.

   c If, instead, the two largest values of x (i.e. 11 and 11) are changed to −9 and b, so that the mean of x decreases by 1, find the value of b and investigate the effect that this has on the mode, the median and the shape of the curve.

14 A histogram is drawn to illustrate a set of continuous data whose mean and median are equal. Make sketches of the different types of curve that could be drawn to represent the shape of the histogram.

15 Students' marks in a Biology examination are shown by percentage in the following table.

Marks (%)	20–	30–	40–	50–	60–	70–	80–90
Frequency (%)	5	10	20	30	20	10	5

   a Without drawing an accurate histogram, describe the shape of the set of marks. What does the shape suggest about the values of the mean, the median and the mode?

   b Information is provided about the marks in examinations in two other subjects:

      Chemistry: mode > median > mean   Physics: mean > median > mode

      Sketch a curve to show the shape of the distribution of marks in each of these exams.

Checklist of learning and understanding

- Measures of central tendency are the mode, the mean and the median.
- For ungrouped data, the mode is the most frequently occurring value.
- For grouped data, the modal class has the highest frequency density and the greatest height column in a histogram.
- For ungrouped data, $\bar{x} = \dfrac{\Sigma x}{n}$.
- For grouped data, $\bar{x} = \dfrac{\Sigma xf}{\Sigma f}$ or $\dfrac{\Sigma fx}{\Sigma f}$.
- The formulae for ungrouped and grouped coded data can be summarised by:

$\bar{x} = \text{mean}(x - b) + b$

$\bar{x} = \dfrac{1}{a} \times [\text{mean}(ax - b) + b]$

- For ungrouped coded data:

$\bar{x} = \dfrac{\Sigma(x - b)}{n} + b$

$\bar{x} = \dfrac{1}{a}\left[\dfrac{\Sigma(ax - b)}{n} + b\right]$

- For grouped coded data:

$\bar{x} = \dfrac{\Sigma(x - b)f}{\Sigma f} + b$

$\bar{x} = \dfrac{1}{a}\left[\dfrac{\Sigma(ax - b)f}{\Sigma f} + b\right]$

- For ungrouped data, the median is at the $\left(\dfrac{n+1}{2}\right)$th value.
- For grouped data, we estimate the median to be at the $\dfrac{n}{2}$th value on a cumulative frequency graph.

Chapter 2: Measures of central tendency

END-OF-CHAPTER REVIEW EXERCISE 2

1 For each of the following sets of data, decide whether you would expect the mean to be less than, equal to or greater than the median and the mode.

   a The ages of patients receiving long-term care at a hospital. [1]

   b The numbers of goals scored in football matches. [1]

   c The heights of adults living in a particular city. [1]

2 The mean mass of 13 textbooks is 875 grams, and n novels have a total mass of 13 706 grams. Find the mean mass of a novel, given that the textbooks and novels together have a mean mass of 716.6 grams. [3]

3 Nine values are 7, 13, 28, 36, 13, 29, 31, 13 and x.

   a Write down the name and the value of the measure of central tendency that can be found without knowing the value of x. [1]

   b If it is known that x is greater than 40, which other measure of central tendency can be found and what is its value? [1]

   c If the remaining measure of central tendency is 25, find the value of x. [2]

4 For the data shown in the following table, x has a mean of 7.15.

x	3	6	10	15
Frequency	a	b	c	d

   a Find the mean value of y given in the following table. [1]

y	11	14	18	23
Frequency	a	b	c	d

   b Find a calculated estimate of the mean value of z given in the following table. [2]

z	2–	8–	14–	24–34
Frequency	a	b	c	d

5 The table below shows the number of books read last month by a group of children.

No. books	2	3	4	5
No. children	3	8	15	q

   a If the mean number of books read is exactly 3.75, find the value of q. [2]

   b Find the greatest possible value of q if:

      i the modal number of books read is 4 [1]

      ii the median number of books read is 4. [1]

6 The following table gives the heights, to the nearest 5 cm, of a group of people.

Heights (cm)	120–135	140–150	155–160	165–170	175–185
No. people	30	p	12	16	21

Given that the modal class is 140–150 cm, find the least possible value of p. [3]

7   The following histogram illustrates the masses, m kilograms, of the 216 sales of hay that a farmer made to customers last year.

    a   Show that a calculated estimate of the mean is equal to the median. [4]

    b   Estimate the price per kilogram at which the hay was sold, given that these sales generated exactly $1944. Why is it possible that none of the customers actually paid this amount per kilogram for the hay? [4]

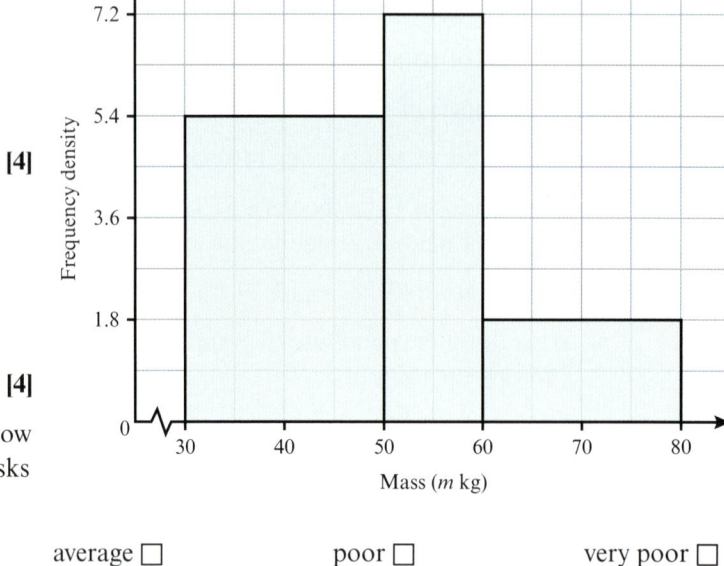

8   An internet service provider wants to know how customers rate its services. A questionnaire asks customers to tick one of the following boxes.

    excellent ☐   good ☐   average ☐   poor ☐   very poor ☐

    a   How might the company benefit from knowing each of the available average responses of its customers? [2]

    b   What additional benefit could the company obtain by using the following set of tick boxes instead?

    excellent = 5 ☐   good = 4 ☐   average = 3 ☐   poor = 2 ☐   very poor = 1 ☐ [2]

9   The numbers of items returned to the electrical department of a store on each of 100 consecutive days are given in the following table.

No. items	0	1	2	3	4	5	6–p
No. days	49	16	10	9	7	5	4

    a   Write down the median. [1]

    b   Is the mode a good value to use as the average in this case? Give a reason for your answer. [1]

    c   Find the value of p, given that a calculated estimate of the mean is 1.5. [3]

    d   Sketch a curve that shows the shape of this set of data, and mark onto it the relative positions of the mode, the mean and the median. [2]

10 As part of a data collection exercise, members of a certain school year group were asked how long they spent on their Mathematics homework during one particular week. The times are given to the nearest 0.1 hour. The results are displayed in the following table.

Time spent (t hours)	$0.1 \leq t \leq 0.5$	$0.6 \leq t \leq 1.0$	$1.1 \leq t \leq 2.0$	$2.1 \leq t \leq 3.0$	$3.1 \leq t \leq 4.5$
Frequency	11	15	18	30	21

    i   Draw, on graph paper, a histogram to illustrate this information. [5]

    ii Calculate an estimate of the mean time spent on their Mathematics homework by members of this year group. [3]

    Cambridge International AS & A Level Mathematics 9709 Paper 6 Q5 June 2008

11 For 150 values of x, it is given that $\Sigma(x-1)+\Sigma(x-4) = 4170$. Find $\bar{x}$. [3]

12 On Monday, a teacher asked eight students to write down a number, which is denoted by x. On Tuesday, when one of these students was absent, they asked them to add 1 to yesterday's number and write it down. Find the number written down on Monday by the student who was absent on Tuesday, given that $\bar{x} = 30\frac{1}{4}$, and that the mean of Monday's and Tuesday's numbers combined was $27\frac{1}{3}$. [3]

13 A delivery of 150 boxes, each containing 20 items, is made to a retailer. The numbers of damaged items in the boxes are shown in the following table.

No. damaged items	0	1	2	3	4	5	6 or more
No. boxes (f)	100	10	10	10	10	10	0

a Find the mode, the mean and the median number of damaged items. [3]

b Which of the three measures of central tendency would be the most appropriate to use as the average in this case? Explain why using the other two measures could be misleading. [2]

14 The monthly salaries, w dollars, of 10 women are such that $\Sigma(w-3000) = -200$.

The monthly salaries, m dollars, of 20 men are such that $\Sigma(m-4000) = 120$.

a Find the difference between the mean monthly salary of the women and the mean monthly salary of the men. [3]

b Find the mean monthly salary of all the women and men together. [3]

15 For 90 values of x and 64 values of y, it is given that $\Sigma(x-1) = 72.9$ and $\Sigma(y+1) = 201.6$. Find the mean value of all the values of x and y combined. [3]

Chapter 3
Measures of variation

In this chapter you will learn how to:

- find and use different measures of variation
- use a cumulative frequency graph to estimate medians, quartiles and percentiles
- calculate and use the standard deviation of a set of data (including grouped data) either from the data itself or from given totals Σx and Σx^2, or coded totals $\Sigma(x-b)$ and $\Sigma(x-b)^2$ and use such totals in solving problems that may involve up to two datasets.

Chapter 3: Measures of variation

PREREQUISITE KNOWLEDGE

Where it comes from	What you should be able to do	Check your skills
IGCSE / O Level Mathematics	Accurately label and read from an axis, using a given scale.	1 The numbers 2 and 18 are marked on an axis 20 cm apart. How far apart are the numbers 4.5 and 17.3 on this axis?
	Substitute into and manipulate algebraic formulae containing squares and square roots.	2 If $y = \sqrt{\dfrac{x^2}{a} - \left(\dfrac{b}{a}\right)^2}$, find the positive value of: a y when $x = 13$, $a = 4$ and $b = \sqrt{352}$ b x when $y = 12$, $a = 5$ and $b = 11$.

How do we best summarise a set of data?

A measure of central tendency alone does not describe or summarise a set of data fully. Although it may tell us the location of the more central values or the most common values, it tells us nothing about how widely spread out the values are. Two sets of data can have the same mean, median or mode, yet they can be completely different. A better description of a set of data is given by a measure of central tendency and a measure of **variation**. Variation is also known as *spread* or *dispersion*.

Consider the runs scored by two batters in their past eight cricket matches, which are given in the following table.

| Batter A | 25 | 30 | 31 | 26 | 31 | 28 | 29 | 24 | Total: 224 |
| Batter B | 2 | 70 | 1 | 0 | 43 | 1 | 104 | 3 | Total: 224 |

The mean number of runs scored by A and by B is the same; namely, $224 \div 8 = 28$. However, the patterns of the number of runs are clearly very different. The numbers for batter A are quite consistent, whereas the numbers for batter B are quite varied. This consistency (or lack of it) can be indicated by a measure of variation, which shows how spread out a set of data values are.

Three commonly used measures of variation are the **range**, **interquartile range** and **standard deviation**.

3.1 The range

As you will recall, the range is the numerical difference between the largest and smallest values in a set of data. One advantage of using the range is that it is easy to calculate. However, it does not take the more central values into account but uses only the most extreme values. It is often more informative to state the minimum and maximum values rather than the difference between them.

For example, in a test for which the lowest mark is 6 and the highest mark is 19, the range is $19 - 6 = 13$.

For grouped data, we can find a minimum and maximum possible range, using the lower and upper boundary values of the data.

WORKED EXAMPLE 3.1

To the nearest centimetre, the tallest and shortest pupils in a class are 169cm and 150cm.

Find the least and greatest possible range of the students' heights.

Answer

Least possible range = 168.5 − 150.5 .... The intervals in which the given heights, h, lie are
= 18 cm $168.5 \leqslant h < 169.5$ cm and $149.5 \leqslant h < 150.5$ cm.
Greatest possible range = 169.5 − 149.5
= 20 cm

3.2 The interquartile range and percentiles

The lower **quartile**, median and upper quartile, as you will recall, divide the values in a dataset into four parts, with an equal number of values in each part.

These three measures are commonly abbreviated by:

- Q_1 for the lower quartile
- Q_2 for the median (or middle quartile)
- Q_3 for the upper quartile.

The interquartile range is the numerical difference between the upper quartile and the lower quartile, and gives the range of the middle half (50%) of the values, as shown in the following diagram.

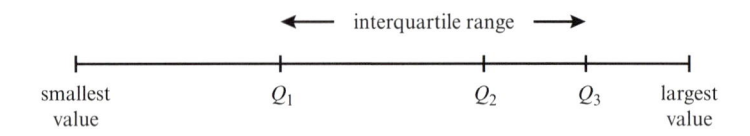

KEY POINT 3.1

Interquartile range = upper quartile − lower quartile or IQR = $Q_3 - Q_1$.

The interquartile range is often preferred to the range because it gives a measure of how varied the more central values are. It is relatively unaffected by extreme values, also called **outliers**, and can be found even when the exact values of these are not known.

Ungrouped data

The positions of the lower and upper quartiles depend on whether there are an odd or even number of values in the set of data. One method that we can use to find the quartiles is as follows.

For an <u>even number</u> of ordered values: we split the data into a lower half and an upper half. Then Q_1 and Q_3 are the medians of the lower half and upper half, respectively.

For an <u>odd number</u> of ordered values: we split the data into a lower half and an upper half at the median, which we then discard. Again, Q_1 and Q_3 are the medians of the lower half and upper half, respectively.

TIP

In a set of ungrouped data, the median is always at the $\left(\frac{n+1}{2}\right)$th value. However, it is advisable to find the quartiles by inspection rather than by memorising formulae for their positions.

Chapter 3: Measures of variation

WORKED EXAMPLE 3.2

Find the interquartile range of the eight ordered values 2, 5, 9, 13, 29, 33, 49 and 55.

Answer

1st	2nd	3rd	4th	5th	6th	7th	8th
2	5	9	13	29	33	49	55

Q_1 between 5 and 9; Q_2 between 13 and 29; Q_3 between 33 and 49.

The ordered values are shown, with their positions indicated.

$IQR = Q_3 - Q_1$
$= \dfrac{33+49}{2} - \dfrac{5+9}{2}$
$= 41 - 7$
$= 34$

WORKED EXAMPLE 3.3

Find the interquartile range of the seven values 69, 17, 43, 6, 73, 77 and 39.

Answer

1st	2nd	3rd	4th	5th	6th	7th
6	17	39	43	69	73	77

$Q_1 = 17$, $Q_2 = 43$, $Q_3 = 73$.

The ordered values and their positions are shown.

$IQR = Q_3 - Q_1$
$= 73 - 17$
$= 56$

WORKED EXAMPLE 3.4

Find the interquartile range of the 13 grouped values shown in the following stem-and-leaf diagram.

```
14 | 2 2 4•8 9      Key: 14 | 2
15 | 1 3 5 6 7•9    represents 142
16 | 5 8
```

REWIND

We studied stem-and-leaf diagrams in Chapter 1, Section 1.2.

Answer

Q_2 is at the $\left(\dfrac{13+1}{2}\right)$th = 7th value,

which is $\boxed{153}$.

We identify the median as 153, which we now discard. This leaves a lower half (142 to 151) and an upper half (155 to 168), with six values in each.

$Q_1 = \dfrac{144+148}{2} = 146$

$Q_3 = \dfrac{157+159}{2} = 158$

The median of a group of six values is the 3.5th value. These are marked by red dots in the stem-and-leaf diagram.

$IQR = 158 - 146$
$= 12$

EXPLORE 3.1

In this activity, you will investigate the value of the median in relation to the smallest and largest values in a set of data, and also in relation to the lower and upper quartiles.

For each ordered set of data, A to D, write down these five values: the smallest value; the lower quartile; the median; the upper quartile; and the largest value.

Set A: 2, 2, 3, 11, 11, 21, 22.

Set B: 6, 6, 6, 11, 13, 17, 19, 20.

Set C: 9, 15, 28, 32, 35, 49.

Set D: 5, 7, 9, 10, 11, 12, 12, 16, 17.

It may be useful to mark the five values for each dataset on a number line.

Use your results to decide which of the following statements are always, sometimes or never true.

1. The median is mid-way between the smallest and largest values.
2. The median is mid-way between the lower and upper quartiles.
3. The interquartile range is equal to exactly half of the range.
4. $Q_3 - Q_2 > Q_2 - Q_1$

Grouped data

We can use a cumulative frequency graph to estimate values in any position in a set of data. This includes the lower quartile, the upper quartile and any chosen percentile.

REWIND

We estimated the median from a cumulative frequency graph in Chapter 1, Section 1.4.

KEY POINT 3.2

For grouped data with total frequency $n = \Sigma f$, the positions of the quartiles are shown in the following table.

Quartile	lower (Q_1)	median (Q_2)	upper (Q_3)
Position	$\dfrac{n}{4}$ or $\dfrac{1}{4}\Sigma f$	$\dfrac{n}{2}$ or $\dfrac{1}{2}\Sigma f$	$\dfrac{3n}{4}$ or $\dfrac{3}{4}\Sigma f$

The nth percentile is the value that is $n\%$ of the way through a set of data.

Q_1, Q_2 and Q_3 are the 25th, 50th and 75th percentiles, respectively.

In an ordered dataset with, say, 320 values, Q_1, Q_2 and Q_3 are at the 80th, 160th and 240th values, and the 90th percentile is at the $(0.90 \times 320) = 288$th value.

The range of the middle 80% of a dataset is the difference between the 10th and 90th percentiles.

WORKED EXAMPLE 3.5

The following graph illustrates the times, in minutes, taken by 500 people to complete a task. Use the graph to find an estimate of:

a the greatest possible range **b** the interquartile range **c** the 95th percentile.

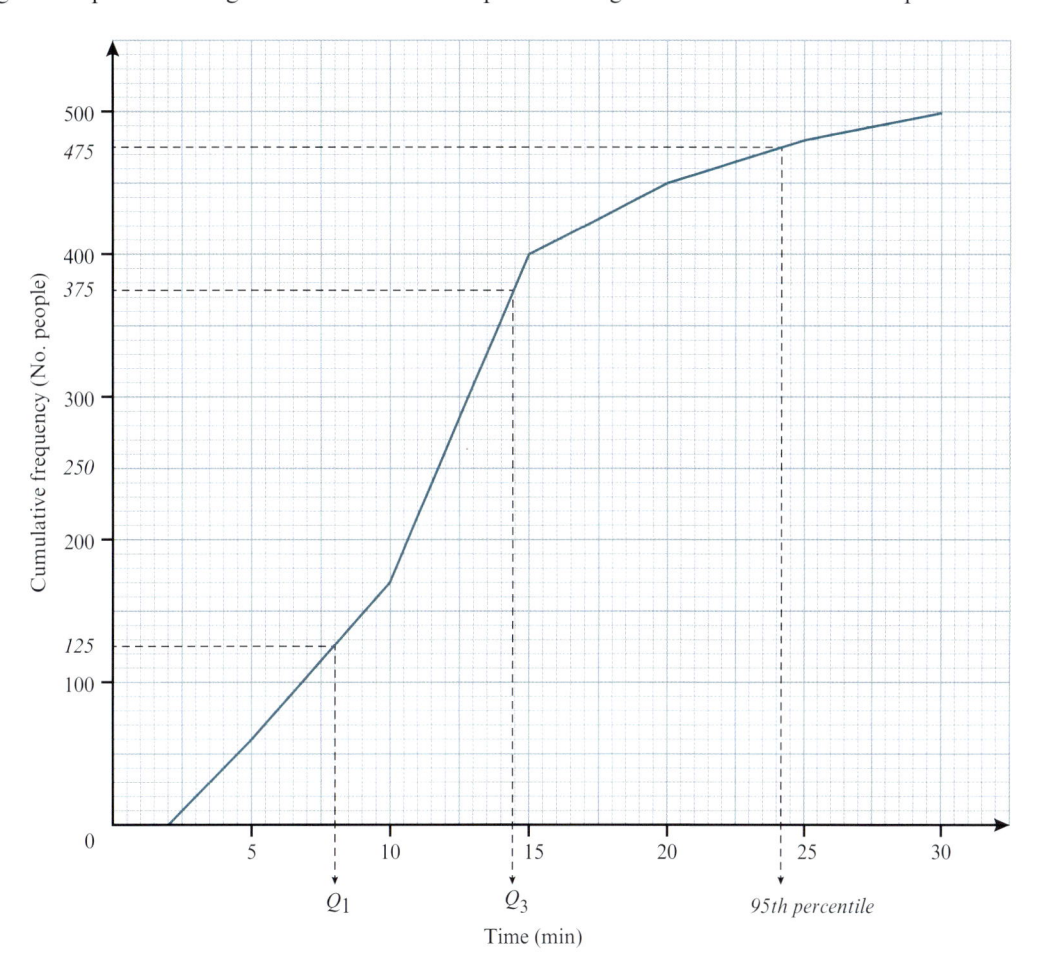

[Cumulative frequency graph showing times in minutes (x-axis, 0 to 30) versus cumulative frequency (No. people, y-axis, 0 to 500). Dashed lines indicate $Q_1 \approx 8$, $Q_3 \approx 14.5$, and the 95th percentile ≈ 24.]

Answer

a $30 - 2 = 28$ min — The greatest possible range is equal to the width of the polygon.

b $Q_1 \approx 8.0$ min — We locate the quartiles, then estimate their values by reading from the graph.
$Q_3 \approx 14.5$ min
$\text{IQR} = Q_3 - Q_1$
$\approx 14.5 - 8.0$
$= 6.5$ min

Lower quartile: $\dfrac{n}{4} = \dfrac{500}{4} = 125$th value

Upper quartile: $\dfrac{3n}{4} = \dfrac{3 \times 500}{4} = 375$th value

c ≈ 24.0 min — The 95th percentile is at the $(0.95 \times 500) = 475$th value.

Box-and-whisker diagrams

A box-and-whisker diagram (or box plot) is a graphical representation of data, showing some of its key features. These features are its smallest and largest values, its lower and upper quartiles, and its median.

If drawn by hand, the diagram is best drawn on graph paper and must include a scale.

It takes the form shown in the following diagram, which shows some features of a dataset denoted by x.

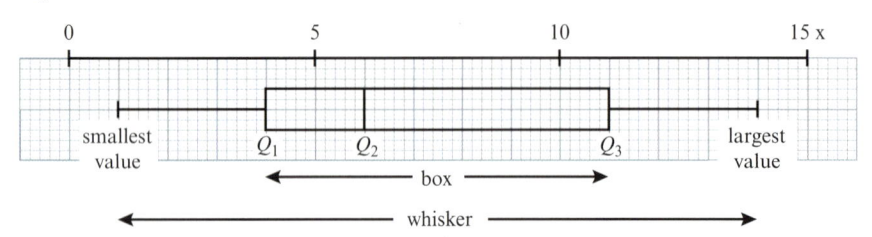

> **TIP**
>
> The whisker (which shows the range) is not drawn through the box (which shows the interquartile range).
>
> Items and, where appropriate, units such as 'Length (cm)' and 'Mass (kg)' must be indicated on the diagram.

Key features of the data for x represented in the box-and-whisker diagram are:

Median: $Q_2 = 6$
Range $= 14 - 1 = 13$
IQR $= Q_3 - Q_1 = 11 - 4 = 7$

The following box-and-whisker diagram is a representation of a dataset denoted by y, drawn using the same scale as the previous diagram.

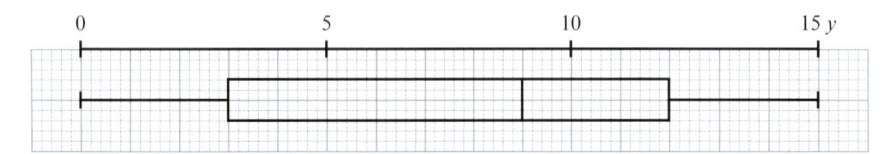

The following table shows a measure of central tendency and two measures of variation for each of x and y, and we can use these to make comparisons.

	Median	Range	IQR
Dataset x	6	13	7
Dataset y	9	$15 - 0 = 15$	$12 - 3 = 9$

By comparing medians, values of x are, on average, less than values of y.

By comparing ranges and interquartile ranges, values of y are more varied than values of x.

We can assess the skewness of a set of data using the quartiles in a box-and-whisker diagram.

In the previous box-and-whisker diagram for x, $Q_3 - Q_2 > Q_2 - Q_1$, and the longer tail of the curve drawn over a bar chart would be to the side of the larger values. This means that the data for x is positively skewed.

In the previous box-and-whisker diagram for y, $Q_2 - Q_1 > Q_3 - Q_2$, and the longer tail of the curve would be to the side of the smaller values. This means that the data for y is negatively skewed.

A reasonably symmetrical set of data would have $Q_3 - Q_2 \approx Q_2 - Q_1$.

> **REWIND**
>
> We looked briefly at positively and negatively skewed sets of data in Chapter 2, Section 2.3.

Chapter 3: Measures of variation

EXERCISE 3A

1. Find the range and the interquartile range of the following sets of data.
   a 5, 8, 13, 17, 22, 25, 30
   b 7, 13, 21, 2, 37, 28, 17, 11, 2
   c 42, 47, 39, 51, 73, 18, 83, 29, 41, 64
   d 113, 97, 36, 81, 49, 41, 20, 66, 28, 32, 17, 107
   e 4.6, 0, −2.6, 0.8, −1.9, −3.3, 5.2, −3.2

2. a Find the range and the interquartile range of the dataset represented in the following box plot.

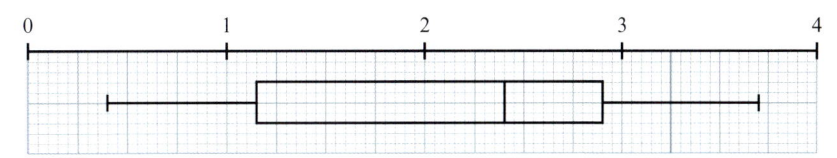

   b What type of skewness would you expect this set of data to have?

3. The following stem-and-leaf diagram shows the marks out of 50 obtained by 15 students in a Science test.

   ```
   0 | 9           Key: 2|5
   1 | 4           represents a
   2 | 5 8         mark of 25
   3 | 0 1 1 7 9   out of 50
   4 | 3 5 6 8
   5 | 0 0
   ```

   a Find the range and interquartile range of the marks.
   b Illustrate the data in a box-and-whisker diagram on graph paper and include a scale.
   c For this set of data, express Q_3 in terms of Q_1 and Q_2.

4. The numbers of fouls made in eight hockey matches and in eight football matches played at the weekend are shown in the following back-to-back stem-and-leaf diagram.

   ```
   Hockey (8) |   | Football (8)    Key: 6|1|8
         4 2 1 | 1 | 0               represents 16 fouls
         9 8 6 | 1 | 8 9 9           in hockey and 18
           4 1 | 2 | 1 2 3 3         fouls in football
   ```

   a Is it true to say that the numbers of fouls in the two sports are equally varied? Explain your answer.
   b Draw two box-and-whisker diagrams using the same scales. Write a sentence to compare the numbers of fouls committed in the two sports.

5. Rishi and Daisy take the same seven tests in Mathematics, and both students' marks improve on successive tests. Their percentage marks are as follows.

   Rishi's marks

   15 24 28 33 39 42 50

   Daisy's marks

   51 65 69 72 78 83 86

   a Explain why it would not be useful to use the range or the interquartile range alone as measures for comparing the marks of the two students.
   b Name two measures that could be used together to give a meaningful comparison of the two students' marks.

6 The following table shows the maximum speeds, s km/h, of some vintage cars.

   a On the same sheet of graph paper, construct a cumulative frequency polygon and a box-and-whisker diagram to illustrate the data.

   b Use your box-and-whisker diagram to assess what type of skewness the data have.

Maximum speed (s km/h)	No. vintage cars (f)
$s < 35$	0
$35 \leq s < 40$	20
$40 \leq s < 45$	65
$45 \leq s < 50$	110
$50 \leq s < 55$	27
$55 \leq s < 70$	13
$70 \leq s < 75$	5

7 Twenty adults are selected at random, and each is asked to state the number of trips abroad that they have made. The results are shown in the following back-to-back stem-and-leaf diagram.

   a i Draw box-and-whisker diagrams, using the same scales for males and for females.

      ii Interpret the key features of the data represented in your diagrams and compare the data for the two groups of adults.

   b In a summary of the data, a student writes, 'The females have visited more countries than the males.' Is this statement justified? Give a reason to support your answer.

```
     Males (9) |   | Females (11)    Key: 3 | 1 | 1
    8 3 2 0 0 0 | 0 | 3 4 5          represents 13
            5 3 | 1 | 1 1 2 3 6      trips for a male
                | 2 | 0 1 2          and 11 trips for
              9 | 3 |                a female
```

8 The resistances, in ohms (Ω), of 100 conductors are represented in the following graph.

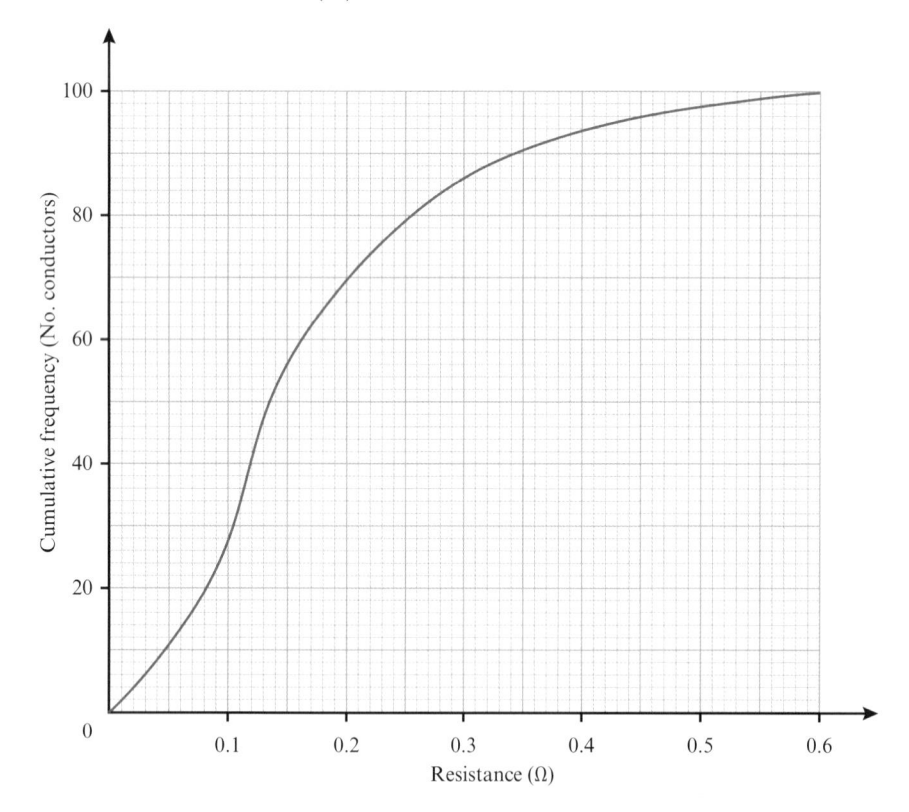

Find, to an appropriate degree of accuracy, an estimate of:

   a the interquartile range
   b the 90th percentile
   c the percentile that is equal to 0.192Ω
   d the range of the middle 40% of the resistances.

9 The areas, in cm², of some circuit boards are represented in the following graph.

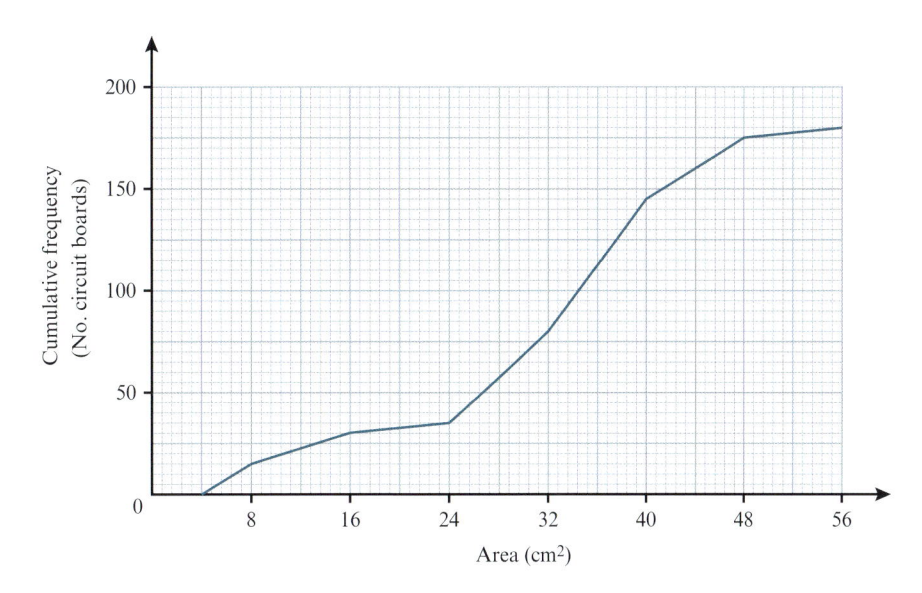

a State the greatest possible range of the data.

b Construct a box-and-whisker diagram to illustrate the data.

c Find the range of the middle 60% of the areas.

d An outlier is an extreme value that is more than 1.5 times the interquartile range above the upper quartile or more than 1.5 times the interquartile range below the lower quartile. Find the areas that define the outliers in this set of data, and estimate how many there are. How accurate is your answer?

10 A company manufactures right-angled brackets for use in the construction industry. A sample of brackets are measured, and the number of degrees by which their angles deviate from a right angle are summarised in the following table.

a Draw a cumulative frequency polygon to illustrate these deviations.

b Estimate the median and the interquartile range of the bracket angles, giving both answers correct to 1 decimal place.

c A bracket is considered unsuitable for use if its angle deviates from a right angle by more than 1.2°. Estimate what percentage of this sample is unsuitable for use, giving your answer correct to the nearest integer.

Deviation from 90° (d)	No. brackets (f)
$d < -1.5$	0
$-1.5 \leq d < -1.0$	24
$-1.0 \leq d < -0.5$	46
$-0.5 \leq d < 0.0$	61
$0.0 \leq d < 0.5$	34
$0.5 \leq d < 1.0$	34
$1.0 \leq d < 1.5$	20
$1.5 \leq d < 2.5$	17

11 The following table shows the cumulative frequencies for values of x.

x	< 0	< 10	< 15	< 25	< 30	< 40
cf	0	12	30	90	102	120

Without drawing a cumulative frequency graph, find:

a the interquartile range

b the 85th percentile.

M **12** Fifty 10-gram samples of a particular type of mushroom are collected by volunteers at a university and tested. The following table shows the mass of toxins, in hundredths of a gram, in these samples.

Mass (/0.01g)	0–	4–	11–	17–	20–30
No. 10 g samples	2	19	23	3	3

a Draw a cumulative frequency curve to illustrate the data.

b Use your curve to estimate, correct to 2 decimal places:

   i the interquartile range

   ii the range of the middle 80%.

c It was found that toxins made up between 0.75% and 2.25% of the mass of n of these samples. Use your curve to estimate the value of n.

d Make an assessment of the variation in the percentage of toxic material in these samples. Can you suggest any possible reasons for such variation?

M **13** A 9-year study was carried out on the pollutants released when biomass fuels are used for cooking. Researchers offered nearly 1000 people living in 12 villages in southern China access to clean biogas and to improved kitchen ventilation. Some people took advantage of neither; some changed to clean fuels; some improved their kitchen ventilation; and some did both. The following diagram shows data on the concentrations of nitrogen dioxide in these people's homes at the end of the study.

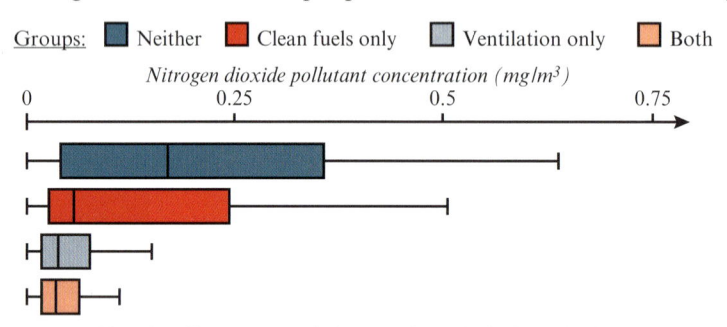

Study the data represented in the diagram and then write a brief analysis that summarises the results of this part of the study.

DID YOU KNOW?

The study of human physical growth, auxology, is a multidisciplinary science involving genetics, health sciences, sociology and economics, among others.

Exceptional height variation in populations that share a genetic background and environmental factors is sometimes due to dwarfism or gigantism, which are medical conditions caused by specific genes or abnormalities in the production of hormones. In regions of poverty or warfare, environmental factors, such as chronic malnutrition during childhood, may result in delayed growth and/or significant reductions in adult stature even without the presence of these medical conditions.

At the time of their meeting in London in 2014, Chandra Bahadur Dangi (at 54.6 cm) and Sultan Kosen (at 254.3 cm) were the shortest and tallest adults in the world.

Chapter 3: Measures of variation

> **EXPLORE 3.2**
>
> The interquartile range is based around the median. In this exploration, we investigate a possible way to define variation based on the mean.
>
> Choose a set of five numbers with a mean of 10.
>
> The *deviation* of a number tells us how far and to which side of the mean it is. Numbers greater than the mean have a positive deviation, whereas numbers less than the mean have a negative deviation, as indicated in the following diagram.
>
> <----- negative deviation | mean | positive deviation ----->
>
> Find the deviation of each of your five numbers and then calculate the mean deviation. Compare and discuss your results, and investigate other sets of numbers.
>
> Can you predict what the result will be for any set of five numbers with a mean of 10? Can you justify your prediction? What would you expect to happen if you started with any set of five numbers?

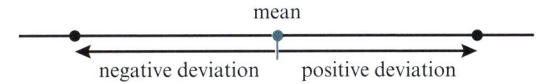

3.3 Variance and standard deviation

In the Explore 3.2 activity, you discovered that the mean deviation is not a useful way of measuring the variation of a dataset because the positive and negative deviations cancel each other out. So, if we want a measure of variation around the mean, we need to ensure that each deviation is positive or zero.

We can do this by calculating the mean *distance* of the data values from the mean, which we call the 'mean absolute deviation from the mean', $\frac{\Sigma|x - \bar{x}|}{n}$.

However, it is hard to calculate this accurately or efficiently for large sets of data and it is difficult to work with algebraically, so this approach is not used in practice.

Alternatively, we can calculate the squared deviation, $(x - \bar{x})^2$ for all data values and find their mean. This is the 'mean squared deviation from the mean', which we call the **variance** of the data.

$$\text{Var}(x) = \frac{\Sigma(x - \bar{x})^2}{n}$$

For measurements and deviations in metres, say, the variance is in m^2. So, to get a measure of variation that is also in metres, we take the square root of the variance, which we call the standard deviation.

Standard deviation of $x = \sqrt{\text{Var}(x)} = \sqrt{\frac{\Sigma(x - \bar{x})^2}{n}}$

This looks no easier to calculate than the 'mean absolute deviation from the mean', however, the formula for variance can be simplified (see appendix at the end of this chapter) to give:

$$\text{Var}(x) = \frac{\Sigma x^2}{n} - \bar{x}^2 = \frac{\Sigma x^2}{n} - \left(\frac{\Sigma x}{n}\right)^2$$

We can find the variance and standard deviation from n, Σx and Σx^2, which are the number of values, their sum and the sum of their squares, respectively. We often use the abbreviation SD(X) to represent the standard deviation of X.

> **TIP**
>
> $|x - \bar{x}|$ means we calculate $x - \bar{x}$ and remove the minus sign if the answer is negative.

> **TIP**
>
> To find Σx^2, we add up the squares of the data values. A common error is to add up the data values and then square the answer but this would be written as $(\Sigma x)^2$ instead.

> **TIP**
>
> To find each value of $x^2 f$ (which is the same as fx^2), we can either multiply x^2 by f or we can multiply x by xf. If we multiply x by f and then square the answer, we will obtain $(xf)^2 = x^2 f^2$, which is not required.

A low standard deviation indicates that most values are close to the mean, whereas a high standard deviation indicates that the values are widely spread out from the mean.

Consider a drinks machine that is supposed to dispense 400 ml of coffee per cup. We would expect some variation in the amount dispensed, yet if the standard deviation is high then some customers are likely to feel cheated and some risk being injured because of their overflowing cups!

> **KEY POINT 3.3**
>
> For ungrouped data:
>
> $$\text{Standard deviation} = \sqrt{\text{Variance}} = \sqrt{\frac{\Sigma(x-\bar{x})^2}{n}} = \sqrt{\frac{\Sigma x^2}{n} - \bar{x}^2}, \text{ where } \bar{x} = \frac{\Sigma x}{n}.$$
>
> For grouped data:
>
> $$\text{Standard deviation} = \sqrt{\text{Variance}} = \sqrt{\frac{\Sigma(x-\bar{x})^2 f}{\Sigma f}} = \sqrt{\frac{\Sigma x^2 f}{\Sigma f} - \bar{x}^2}, \text{ where } \bar{x} = \frac{\Sigma xf}{\Sigma f}.$$

> **TIP**
>
> We can remember the formula for variance as 'mean of the squares minus square of the mean'.

WORKED EXAMPLE 3.6

For the set of five numbers 3, 9, 15, 24 and 29, find:

a the standard deviation

b which of the five numbers are more than one standard deviation from the mean.

Answer

a $\text{Variance} = \frac{\Sigma x^2}{n} - \left(\frac{\Sigma x}{n}\right)^2$

$= \frac{3^2 + 9^2 + 15^2 + 24^2 + 29^2}{5} - \left(\frac{3+9+15+24+29}{5}\right)^2$

$= \frac{1732}{5} - \left(\frac{80}{5}\right)^2$

$= 346.4 - 16^2$

$= 90.4$

Standard deviation $= \sqrt{90.4}$

$= 9.51$

We subtract the square of the mean from the mean of the squares to find the variance.

We take the square root of the variance to find the standard deviation, correct to 3 significant figures.

b $16 - 9.51 = 6.49$

$16 + 9.51 = 25.51$

The numbers are 3 and 29.

We find the values that are 9.51 below and 9.51 above, using the mean of 16.

Identify which of the five numbers are outside the range $6.49 \leqslant \text{number} \leqslant 25.51$.

Chapter 3: Measures of variation

WORKED EXAMPLE 3.7

Find the standard deviation of the values of x given in the following table, correct to 3 significant figures.

x	f
12	13
14	28
16	10

Answer

x	f	xf	$x^2 f = x \times xf$
12	13	156	$12 \times 156 = 1872$
14	28	392	$14 \times 392 = 5488$
16	10	160	$16 \times 160 = 2560$
	$\Sigma f = 51$	$\Sigma xf = 708$	$\Sigma x^2 f = 9920$

The table shown opposite is an extended frequency table that is used to find Σf, Σxf and $\Sigma x^2 f$, which are needed to calculate the standard deviation.

$$SD(x) = \sqrt{\frac{\Sigma x^2 f}{\Sigma f} - \bar{x}^2}$$

We use the totals 51, 708 and 9920 to find the standard deviation.

$$= \sqrt{\frac{9920}{51} - \left(\frac{708}{51}\right)^2}$$

$$= 1.34$$

> **TIP**
> Note how the values of $x^2 f$ are calculated.

> **TIP**
> Always use the exact value of the mean to calculate variance and standard deviation.

What happens if we use a rounded value for the mean?

Correct to 1 decimal place, the mean in Worked example 3.7 is $708 \div 51 = 13.9$.

If we use $\bar{x} = 13.9$ in our calculation, we obtain $SD(x) = \sqrt{\frac{9920}{51} - 13.9^2} = 1.14$. This is an error of 0.2.

The rounded mean has caused a substantial error (0.2 is about 15% of the correct value 1.34). So, when calculating the variance or standard deviation, always use $\frac{\Sigma x}{n}$ or $\frac{\Sigma xf}{\Sigma f}$, rather than a rounded value for the mean.

When data are grouped, actual values cannot be seen, but we can calculate estimates of the variance and standard deviation. The formulae in Key point 3.3 are used to do this, where x now represents class mid-values and $\bar{x} = \frac{\Sigma xf}{\Sigma f}$ is a calculated estimate of the mean.

WORKED EXAMPLE 3.8

Calculate an estimate of the standard deviation of the heights of the 20 children given in the following table.

Height (metres)	1.2–	1.4–	1.5–1.7
No. children (f)	2	12	6

Answer

Height (m)	1.2–	1.4–	1.5–1.7	
No. children (f)	2	12	6	$\Sigma f = 20$
Mid-value (x)	1.3	1.45	1.6	
xf	2.6	17.4	9.6	$\Sigma xf = 29.6$
$x^2 f$	3.38	25.23	15.36	$\Sigma x^2 f = 43.97$

We extend the frequency table to include class mid-values (x), and to find the totals Σf, Σxf and $\Sigma x^2 f$, as shown in the table opposite.

$$\text{Estimate of standard deviation} = \sqrt{\frac{\Sigma x^2 f}{\Sigma f} - \bar{x}^2}$$

$$= \sqrt{\frac{43.97}{20} - \left(\frac{29.6}{20}\right)^2}$$

$$= \sqrt{0.0081}$$

$$= 0.09 \, \text{m}$$

EXPLORE 3.3

Four students analysed data that they had collected. Their findings are given below.

1. Property prices in a certain area of town have a high standard deviation.

2. The variance of the monthly sales of a particular product last year was high.

3. The standard deviation of students' marks in a particular examination was close to zero.

4. The times taken to perform a new medical procedure have a low variance.

Discuss the students' findings and give a possible description of each of the following.

1. The type of environment and the people purchasing property in this area of town.

2. The type of product being sold.

3. The usefulness of the examination.

4. The efficiency of the teams performing the medical procedures.

Chapter 3: Measures of variation

Although standard deviation is far more commonly used as a measure of variation than the interquartile range, it may not always be ideal because it can be significantly affected by extreme values. The interquartile range may be better, and a box-and-whisker diagram is often much more useful as a visual representation of data than the mean and standard deviation.

Some features of the standard deviation are compared to the interquartile range in the following table.

Advantages of standard deviation	Disadvantages of standard deviation
Much simpler to calculate than the IQR.	Far more affected by extreme values than the IQR.
Data values do not have to be ordered.	Gives greater emphasis to large deviations than to small deviations.
Easier to work with algebraically when doing more advanced work.	
Takes account of all data values.	

EXERCISE 3B

1 Find the mean and the standard deviation for these sets of numbers.

   a 27, 43, 29, 34, 53, 37, 19 and 58.

   b 6.2, −8.5, 7.7, −4.3, 13.5 and −11.9.

2 Last term Abraham sat three tests in each of his science subjects. His raw percentage marks for the tests, in the order they were completed, are listed.

Biology	Chemistry	Physics
21 33 45	41 53 65	51 63 75

   a Calculate the variance of Abraham's marks in each of the three subjects.

   b Comment on the three values obtained in part **a**. Do the same comments apply to Abraham's mean mark for the tests in the three subjects? Justify your answer.

3 The following table shows the number of pets owned by each of 35 families.

No. pets	0	1	2	3	4	5
No. families (f)	6	12	9	4	3	1

   Find the mean and variance of the number of pets.

4 The numbers of cobs produced by 360 maize plants are shown in the following table.

No. cobs	0	1	2	3	4
No. plants (f)	11	75	185	81	8

   a Calculate the mean and the standard deviation.

   b Find the interquartile range and give an example of what it tells us about this dataset that the standard deviation does not tell us.

5 The times spent, in minutes, by 30 girls and by 40 boys on an assignment are detailed in the following table.

Time spent (min)	20–	30–	40–	60–80
No. girls (f)	6	14	7	3
No. boys (f)	15	11	7	7

a For the boys and for the girls, calculate estimates of the mean and standard deviation.

b It is required to make a comparison between the times spent by the two groups.

  i What do the means tell us about the times spent?

  ii Use the standard deviations to compare the times spent by the two groups.

6 The lengths, correct to the nearest centimetre, of 50 rods are given in the following table.

Length (cm)	15–17	18–24	25–29	30–37
No. rods (f)	13	18	11	8

Calculate an estimate of the standard deviation of the lengths.

7 For the dataset denoted by x in the following table, k is a constant.

x	15	16	17	18	19	20
f	$2k$	$k+5$	$k-3$	10	8	3

Find the value of k and calculate the variance of x, given that $\bar{x} = 17$.

8 The following table illustrates the heights, in centimetres, of 150 children.

Height (cm)	No. children (f)
140 up to 144	a
144 up to 150	b
150 up to 160	69
160 up to 165	28

a Given that a calculated estimate of the mean height is exactly 153.14 cm, show that $142a + 147b = 7726$, and evaluate a and b.

b Calculate an estimate of the standard deviation of the heights.

PS 9 Kristina plans to raise money for charity. Her plan is to walk 217 km in 7 days so that she walks $k + 2n - n^2$ km on the nth day. Find the standard deviation of the daily distances she plans to walk, and compare this with the interquartile range.

PS 10 The mass of waste produced by a school during its three 13-week terms is given in tonnes, correct to 2 decimal places, in the following table.

Mass of waste (tonnes)	0.15–0.29	0.30–0.86	0.87–1.35	1.36–2.00
No. weeks (f)	5	8	20	6

a Calculate estimates of the mean and standard deviation of the mass of waste produced per week, giving both answers correct to 2 decimal places

b No waste is produced in the 13 weeks of the year that the school is closed. If this additional data is included in the calculations, what effect does it have on the mean and on the standard deviation?

11 The ages, in whole numbers of years, of a hotel's 50 staff are given in the following table. Calculated estimates of the mean and variance are 37.32 and 69.1176, respectively.

Age (years)	23–30	31–37	38–45	46–59
No. staff (f)	14	x	y	6

Exactly 1 year after these calculations were made, Gudrun became the 51st staff member and the mean age became exactly 38 years. Find Gudrun's age on the day of her recruitment, and determine what effect this had on the variance of the staff's ages. What assumptions must be made to justify your answers?

12 Refer to the following diagram. In position 1, a 10-metre rod is placed 10 metres from a fixed point, P. Six small discs, A to F, are evenly spaced along the length of the rod. The rod is rotated anti-clockwise about its centre by $\alpha = 30°$ to position 2. The distances from P to the discs are denoted by x.

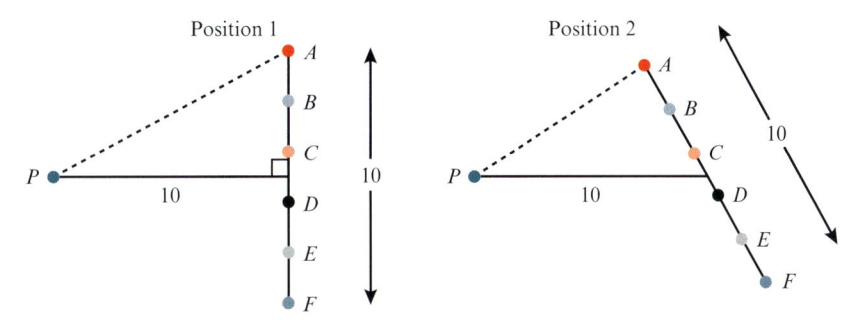

a What effect does the 30° rotation have on values of x? Investigate this by first considering the effect on the average distance from P to the discs.

b Find two values that can be used as measures of the change in the variation of x caused by the rotation.

c Use the values obtained in parts **a** and **b** to summarise the changes in the distances from P to the discs caused by the rotation.

d Can you prove that Σx^2 is constant for all values of α? (Hint: to do this, you need only to show that Σx^2 is constant for $0° < \alpha < 90°$).

EXPLORE 3.4

Twenty adults completed as many laps of a running track as they could manage in 30 minutes. The following table shows how many laps they completed.

Completed laps	4–8	9–13	14–18
No. adults (f)	6	10	4

Two students, Andrea and Billie, were asked to calculate an estimate of the standard deviation. Their working and answers, which you should check carefully, are shown below.

Andrea: $\sqrt{\dfrac{(6^2 \times 6)+(11^2 \times 10)+(16^2 \times 4)}{20} - 10.5^2} = 3.5$

Billie: $\sqrt{\dfrac{(6.5^2 \times 6)+(11.5^2 \times 10)+(16.5^2 \times 4)}{20} - 11^2} = 3.5$

Compare Andrea and Billie's approaches. What have they done differently and why do you think they did so? Is one of their answers better than the other? If so, in what way?

Calculating from totals

For ungrouped data, we calculate variance (Var) and standard deviation (SD) from totals n, Σx and Σx^2.

For grouped data, we calculate using totals Σf, Σxf and $\Sigma x^2 f$.

In both cases, we can rearrange the formula for variance if we wish to evaluate one of the totals.

WORKED EXAMPLE 3.9

Given that $n = 25$, $\Sigma x = 275$ and $\text{Var}(x) = 7$, find Σx^2.

Answer

$$\frac{\Sigma x^2}{25} - \left(\frac{275}{25}\right)^2 = 7$$

$$\Sigma x^2 = 25 \times \left[7 + \left(\frac{275}{25}\right)^2\right]$$

$$= 3200$$

Substitute the given values into the formula for variance, then rearrange the terms to make Σx^2 the subject.

Combined sets of data

In Chapter 2, Section 2.2, sets of data were combined by simply considering all of their values together, and we learned how to find the mean. Here, we consider the variation of datasets that have been combined in the same way.

The variance and standard deviation of a combined dataset are calculated from its totals, which are the sums of the totals of the two sets from which it has been made.

The two sets {1, 2, 3, 4} and {4, 5, 6} individually have variances of $1\frac{1}{4}$ and $\frac{2}{3}$. The combined set {1, 2, 3, 4, 4, 5, 6} has a variance of approximately 2.53.

WORKED EXAMPLE 3.10

The heights, x cm, of 10 boys are summarised by $\Sigma x = 1650$ and $\Sigma x^2 = 275490$.

The heights, y cm, of 15 girls are summarised by $\Sigma y = 2370$ and $\Sigma y^2 = 377835$.

Calculate, to 3 significant figures, the standard deviation of the heights of all 25 children together.

Answer

$\Sigma x^2 + \Sigma y^2 = 275490 + 377835 = 653325$

$\Sigma x + \Sigma y = 1650 + 2370 = 4020$

For the 25 children, we find the sum of the squares of their heights and the sum of their heights.

$$\text{Standard deviation} = \sqrt{\frac{653\,325}{25} - \left(\frac{4020}{25}\right)^2}$$

$$= 16.6\,\text{cm}$$

We substitute the three sums into the formula for standard deviation and evaluate this to the required degree of accuracy.

TIP

The variance of two combined datasets x and y is not (in general) equal to $\dfrac{\text{Var}(x) + \text{Var}(y)}{2}$.

In Worked example 3.10, $\text{Var}(\text{boys}) = 324$ and $\text{Var}(\text{girls}) = 225$ but $\text{Var}(\text{boys and girls}) \neq \dfrac{324 + 225}{2}$.

If two sets of data, denoted by x and y, have n_x and n_y values, respectively, then the mean and variance of their combined values are found using the totals $(n_x + n_y)$, $(\Sigma x + \Sigma y)$ and $(\Sigma x^2 + \Sigma y^2)$.

> **KEY POINT 3.4**
>
> The mean of x and y combined is $\dfrac{\Sigma x + \Sigma y}{n_x + n_y}$.
>
> The variance of x and y combined is $\dfrac{\Sigma x^2 + \Sigma y^2}{n_x + n_y} - \left(\dfrac{\Sigma x + \Sigma y}{n_x + n_y}\right)^2$.

We can rearrange the formulae in Key point 3.4 if we wish to find one of the totals involved.

WORKED EXAMPLE 3.11

In an examination, the percentage marks of the 120 boys are denoted by x, and the percentage marks of the 80 girls are denoted by y.

The marks are summarised by the totals $\Sigma x = 7020$, $\Sigma x^2 = 424\,320$ and $\Sigma y^2 = 352\,130$.

Calculate the girls' mean mark, given that the standard deviation for all these students is 10.

Answer

$\dfrac{424\,320 + 352\,130}{120 + 80} - \left(\dfrac{7020 + \Sigma y}{120 + 80}\right)^2 = 10^2$

We substitute the given values into the formula for variance, knowing that this is equal to 10^2.

$3882.25 - \dfrac{(7020 + \Sigma y)^2}{40\,000} = 100$

Then multiply throughout by 40 000.

$155\,290\,000 - (7020 + \Sigma y)^2 = 4\,000\,000$

Then we rearrange to make Σy (the total marks for the girls) the subject.

$(7020 + \Sigma y)^2 = 151\,290\,000$

$\Sigma y = \sqrt{151\,290\,000} - 7020$

We take the positive square root, as we know that all of the y values are non-negative.

$\Sigma y = 5280$

Girls' mean mark $= \dfrac{5280}{80} = 66$

Divide the total marks for the girls by the number of girls.

EXERCISE 3C

1 Given that:

   a $\Sigma v^2 = 5480$, $\Sigma v = 288$ and $n = 64$, find the variance of v.

   b $\Sigma w^2 = 4000$, $\bar{w} = 5.2$ and $n = 36$, find the standard deviation of w.

   c $\Sigma x^2 f = 6120$, $\Sigma f = 40$ and the standard deviation of x is 12, find Σxf.

   d $\Sigma xf = 2800$, $\Sigma f = 50$ and the variance of x is 100, find $\Sigma x^2 f$.

   e $\Sigma t^2 = 193\,144$, $\Sigma t = 2324$ and that the standard deviation of t is 3, find the number of data values of t.

2 A building is occupied by n companies. The number of people employed by these companies is denoted by x. Find the mean number of employees, given that $\Sigma x^2 = 8900$, $\Sigma x = 220$ and that the standard deviation of x is 18.

3 Twenty-five values of p are such that $\Sigma p^2 = 6006$ and $\Sigma p = 388$, and 25 values of q are such that $\Sigma q^2 = 6114$ and $\Sigma q = 387$. Calculate the variance of the 50 values of p and q together.

4 In a class of 30 students, the mean mass of the 14 boys is 63.5 kg and the mean mass of the girls is 57.3 kg. Calculate the mean and standard deviation of the masses of all the students together, given that the sums of the squares of the masses of the boys and girls are 58 444 kg² and 56 222 kg², respectively.

5 The following table shows the front tyre pressure, in psi, of five 4-wheeled vehicles, A to E.

	A	B	C	D	E
Front-left tyre	26	29	30	34	26
Front-right tyre	24	27	31	30	28

a Show that the variance of the pressure in all of these front tyres is 7.65 psi².

b Rear tyre pressures for these five vehicles are denoted by x. Given that $\Sigma x^2 = 7946$ and that the variance of the pressures in all of the front and rear tyres on these five vehicles together is 31.6275 psi², find the mean pressure in all the rear tyres.

6 The totals $\Sigma x^2 = 7931$, $\Sigma x = 397$ and $\Sigma y = 499$ are given by 29 values of x and n values of y. All the values of x and y together have a variance of 52.

a Express Σy^2 in terms of n.

b Find the value of n for which $\Sigma y^2 - \Sigma x^2 = 10$.

PS 7 The five values in a dataset have a sum of 250 and standard deviation of 15. A sixth value is added to the dataset, such that the mean is now 40. Find the variance of the six values in the dataset.

PS 8 A group of 10 friends played a mini-golf competition. Eight of the friends tied for second place, each with a score of 34, and the other two friends tied for first place. Find the winning score, given that the standard deviation of the scores of all 10 friends was 1.2 and that the lowest score in golf wins.

M PS 9 An author has written 15 children's books. The first eight books that she wrote contained between 240 and 250 pages each. The next six books contained between 180 and 190 pages each. Correct to 1 decimal place, the standard deviation of the number of pages in the 15 books together is 31.2.

Show that it is not possible to determine a specific calculated estimate of the number of pages in the author's 15th book.

PS 10 A set of n pieces of data has mean $\bar{x}$ and standard deviation S.

Another set of $2n$ pieces of data has mean $\bar{x}$ and standard deviation $\frac{1}{2}S$.

Find the standard deviation of all these pieces of data together in terms of S.

> ▶▶ **FAST FORWARD**
>
> We will see how standard deviation and probabilities are linked in the normal distribution in Chapter 8, Section 8.2.

Chapter 3: Measures of variation

> **EXPLORE 3.5**
>
> The following table shows three students' marks out of 20 in the same five tests.
>
> | | 1st | 2nd | 3rd | 4th | 5th | |
> |-------|-----|-----|-----|-----|-----|-----|
> | Amber | 12 | 17 | 11 | 9 | 16 | x |
> | Buti | 11 | 16 | 10 | 8 | 15 | $x-1$ |
> | Chen | 15 | 20 | 14 | 12 | 19 | $x+3$ |
>
> Note that Buti's marks are consistently 1 less than Amber's and that Chen's marks are consistently 3 more than Amber's. This is indicated in the last column of the table.
>
> For each student, calculate the variance and standard deviation.
>
> Can you explain your results, and do they apply equally to the range and interquartile range?

Coded data

What effect does addition of a constant to all the values in a dataset have on its variation? And how can we find the variance and standard deviation of the original data from the coded data?

In the Explore 3.5 activity, you discovered that the datasets x, $x-1$ and $x+3$ have identical measures of variation. The effect of adding -1 or $+3$ is to translate the whole set of values, which has no effect on the pattern of spread, as shown in the following diagram. The marks of the three students have the same variance and the same standard deviation.

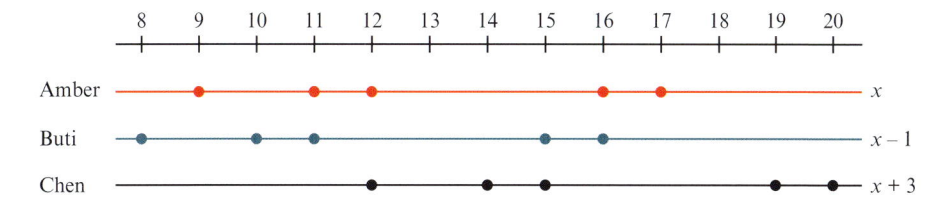

> **REWIND**
>
> We saw in Chapter 2, Section 2.2 that the mean of a set of data can be found from a coded total such as $\Sigma(x-b)$.

> **KEY POINT 3.5**
>
> For ungrouped data: $\dfrac{\Sigma x^2}{n} - \bar{x}^2 = \dfrac{\Sigma(x-b)^2}{n} - \left(\dfrac{\Sigma(x-b)}{n}\right)^2$
>
> For grouped data: $\dfrac{\Sigma x^2 f}{\Sigma f} - \bar{x}^2 = \dfrac{\Sigma(x-b)^2 f}{\Sigma f} - \left(\dfrac{\Sigma(x-b)f}{\Sigma f}\right)^2$
>
> These formulae can be summarised by writing $\text{Var}(x) = \text{Var}(x-b)$.

For two datasets coded as $(x-a)$ and $(x-b)$, we can use the coded totals $\Sigma(x-a)$, $\Sigma(x-a)^2$, $\Sigma(y-b)$ and $\Sigma(y-b)^2$ to find Σx, Σx^2, Σy and Σy^2, from which we can find the variance of the combined set of values of x and y.

WORKED EXAMPLE 3.12

Eight values of x are summarised by the totals $\Sigma(x-10)^2 = 1490$ and $\Sigma(x-10) = 100$.
Twelve values of y are summarised by the totals $\Sigma(y+5)^2 = 5139$ and $\Sigma(y+5) = 234$.
Find the variance of the 20 values of x and y together.

Answer

$\bar{x} = \dfrac{100}{8} + 10 = 22.5$, so $\Sigma x = 8 \times 22.5 = 180$. ...... We find the totals Σx and Σx^2.

$\text{Var}(x) = \text{Var}(x-10) = \dfrac{1490}{8} - \left(\dfrac{100}{8}\right)^2 = 30$.

$\dfrac{\Sigma x^2}{8} - 22.5^2 = 30$, so $\Sigma x^2 = 4290$.

$\bar{y} = \dfrac{234}{12} - 5 = 14.5$, so $\Sigma y = 12 \times 14.5 = 174$. ...... We find the totals Σy and Σy^2.

$\text{Var}(y) = \text{Var}(y+5) = \dfrac{5139}{12} - \left(\dfrac{234}{12}\right)^2 = 48$.

$\dfrac{\Sigma y^2}{12} - 14.5^2 = 48$, so $\Sigma y^2 = 3099$.

$\text{Var}(x \text{ and } y) = \dfrac{\Sigma x^2 + \Sigma y^2}{8+12} - \left(\dfrac{\Sigma x + \Sigma y}{8+12}\right)^2$

$= \dfrac{4290 + 3099}{20} - \left(\dfrac{180 + 174}{20}\right)^2$

$= 56.16$

WORKED EXAMPLE 3.13

It is known that 20 girls each have at least one brother. The number of brothers that they have is denoted by x. Information about the values of $x-1$ is given in the following table.

$x-1$	0	1	2	3	4
No. girls (f)	2	4	8	5	1

Use the coded values to calculate the standard deviation of the number of brothers, to 3 decimal places.

Answer

$x-1$	0	1	2	3	4	
No. girls (f)	2	4	8	5	1	$\Sigma f = 20$
$(x-1)f$	0	4	16	15	4	$\Sigma(x-1)f = 39$
$(x-1)^2 f$	0	4	32	45	16	$\Sigma(x-1)^2 f = 97$

...... We extend the frequency table to find the necessary totals.

$$SD(x) = SD(x-1)$$
$$= \sqrt{\frac{\Sigma(x-1)^2 f}{\Sigma f} - \left(\frac{\Sigma(x-1)f}{\Sigma f}\right)^2}$$
$$= \sqrt{\frac{97}{20} - \left(\frac{39}{20}\right)^2}$$
$$= 1.023$$

> We know that the standard deviation of x and the standard deviation of $(x-1)$ are equal.

EXERCISE 3D

1. Two years ago, the standard deviation of the masses of a group of men and a group of women were 8 kg and 6 kg, respectively. Today, all the men are 5 kg heavier and all the women are 3 kg lighter. Find the standard deviation for each group today.

2. Twenty readings of y are summarised by the totals $\Sigma(y-5)^2 = 890$ and $\Sigma(y-5) = 130$. Find the standard deviation of y.

3. The amounts of rainfall, r mm, at a certain location were recorded on 365 consecutive days and are summarised by $\Sigma(r-3)^2 = 9950$ and $\Sigma(r-3) = 1795.8$. Calculate the mean daily rainfall and the value of Σr^2.

4. Exactly 20 years ago, the mean age of a group of boys was 15.7 years and the sum of the squares of their ages was 16 000. If the sum of the squares of their ages has increased by 8224 in this 20-year period, find the number of boys in the group.

5. Readings from a device, denoted by y, are such that $\Sigma(y-3)^2 = 2775$, $\Sigma y = 105$ and the standard deviation of y is 13. Find the number of readings that were taken.

M 6. Mei measured the heights of her classmates and, after correctly analysing her data, she found the mean and standard deviation to be 163.8 cm and 7.6 cm. Decide whether or not these measures are valid, given the fact that Mei measured all the heights from the end of the tape measure, which is exactly 1.2 cm from the zero mark.

   Explain your answers.

M 7. A transport company runs 21 coaches between two cities every week. In the past, the mean and variance of the journey times were 4 hours 35 minutes and 53.29 minutes^2.

   What would be the mean and standard deviation of the times if all the coaches departed 10 minutes later and arrived 5 minutes earlier than in the past?

   Are there any situations in which achieving this might actually be possible?

PS 8. During a sale, a boy bought six pairs of jeans, each with leg length x cm. He also bought four pairs of pants, each with leg length $(x-2)$ cm. The boy is quite short, so his father removed 4 cm from the length of each trouser leg. Find the variance of the leg lengths after his father made the alterations.

Cambridge International AS & A Level Mathematics: Probability & Statistics 1

P 9 **a** Find the mean and standard deviation of the first seven positive even integers.

b Without using a calculator, write down the mean and standard deviation of the first seven positive odd integers.

c Find an expression in terms of n for the variance of the first n positive even integers. What other measure can be found using this expression?

> **FAST FORWARD**
>
> You will study the variance of linear combinations of random variables in the Probability & Statistics 2 Coursebook, Chapter 3.

10 Each year Upchester United plays against Upchester City in a local derby match. The number of goals scored in a match by United is denoted by u and the number of goals scored in a match by City is denoted by c. The number of goals scored in the past 15 matches are summarised by $\Sigma(u-1)^2 = 25$, $\Sigma(u-1) = 9$, $\Sigma c^2 = 39$ and $\Sigma c = 19$.

a How many goals have been scored altogether in these 15 matches?

b Show that $\Sigma u^2 = 58$.

c Find, correct to 3 decimal places, the variance of the number of goals scored by the two teams together in these 15 matches.

11 Twenty values of x are summarised by $\Sigma(x-1)^2 = 132$ and $\Sigma(x-1) = 44$.

Eighty values of y are summarised by $\Sigma(y+1)^2 = 17\,704$ and $\Sigma(y+1) = 1184$.

a Show that $\Sigma x = 64$ and that $\Sigma x^2 = 240$.

b Calculate the value of Σy and of Σy^2.

c Find the exact variance of the 100 values of x and y combined.

12 The heights, x cm, of 200 boys and the heights, y cm, of 300 girls are summarised by the following totals:

$\Sigma(x-160)^2 = 18\,240$, $\Sigma(x-160) = 1820$, $\Sigma(y-150)^2 = 20\,100$, $\Sigma(y-150) = 2250$.

a Find the mean height of these 500 children.

b By first evaluating Σx^2 and Σy^2, find the variance of the heights of the 500 children, including appropriate units with your answer.

E What effect does multiplication of all the values in a dataset have on its variation? And how can we find the variance and standard deviation of the original values from the coded data?

Consider the total cost of hiring a taxi for which a customer pays a fixed charge of $3 plus $4 per kilometre travelled. Using y for the total cost and x for the distance travelled in kilometres, the cost can be calculated from the equation $y = 4x + 3$. Some example values are shown in the following table.

Distance (x km)	1	2	3
Cost ($\$y$)	7	11	15

$+1 \rightarrow$ (between columns of x)
$+4 \rightarrow$ (between columns of y)

> **REWIND**
>
> We saw in Chapter 2, Section 2.2 that the mean of a set of data can be found from a coded total such as $\Sigma(ax-b)$.

When the distance changes or varies by +1, the cost changes or varies by +4.

Variation in cost is affected by multiplication (×4) but not by addition (+3).

If we consider the graph of $y = 4x + 3$, then it is only the gradient of the line that affects the variation of y. If we increase the x-coordinate of a point on the line by 1, its y-coordinate increases by 4.

Multiplying $x = \{1, 2, 3\}$ by 4 'stretches' the whole set to $\{4, 8, 12\}$, which affects the pattern of spread.

Adding +3 to $\{4, 8, 12\}$ simply translates the whole set to $y = \{7, 11, 15\}$, which has no effect on the pattern of spread.

Journeys of 1, 2 and 3 km, costing \$7, \$11 and \$15 are represented in the following diagram.

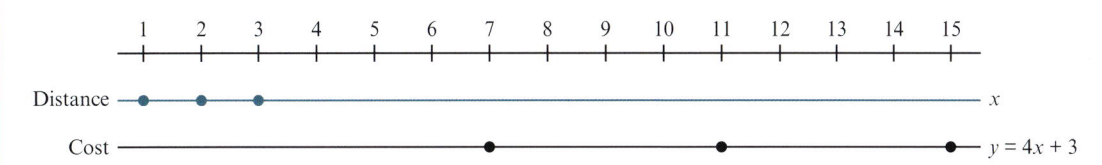

In the diagram, we see that the range of y is 4 times the range of x, so the range of x is $\frac{1}{4}$ times the range of y. For $x = \{1, 2, 3\}$ and $4x + 3 = \{7, 11, 15\}$, you should check to confirm the following results:

$$SD(x) = \frac{1}{4} \times SD(4x+3) \text{ and } Var(x) = \frac{1}{16} \times Var(4x+3).$$

KEY POINT 3.6

For ungrouped data: $\dfrac{\Sigma x^2}{n} - \bar{x}^2 = \dfrac{1}{a^2}\left[\dfrac{\Sigma(ax-b)^2}{n} - \left(\dfrac{\Sigma(ax-b)}{n}\right)^2\right]$

For grouped data: $\dfrac{\Sigma x^2 f}{\Sigma f} - \bar{x}^2 = \dfrac{1}{a^2}\left[\dfrac{\Sigma(ax-b)^2 f}{\Sigma f} - \left(\dfrac{\Sigma(ax-b)f}{\Sigma f}\right)^2\right]$

These formulae can be summarised by writing $Var(x) = \dfrac{1}{a^2} \times Var(ax-b)$ or $Var(ax-b) = a^2 \times Var(x)$.

WORKED EXAMPLE 3.14

The standard deviation of the prices of a selection of brand-name products is \$24. Imitations of these products are all sold at 25% of the brand-name price. Find the variance of the prices of the imitations.

Answer

$Var(0.25x) = 0.25^2 \times Var(x)$

$ = 0.25^2 \times 24^2$

$ = 36$

Denoting the brand-name prices by x and the imitation prices by $0.25x$, we use the fact that $Var(ax) = a^2 \times Var(x)$ to find $Var(0.25x)$.

TIP

The units for variance in this case are 'dollars squared'.

Cambridge International AS & A Level Mathematics: Probability & Statistics 1

WORKED EXAMPLE 3.15

Given that $\Sigma(3x-1)^2 = 9136$, $\Sigma(3x-1) = 53$ and $n = 10$, find the value of Σx^2.

Answer

$3\bar{x} - 1 = \dfrac{53}{10}$

$\bar{x} = \dfrac{1}{3} \times \left(\dfrac{53}{10} + 1\right)$

$= 2.1$

We first find $\bar{x}$, knowing that the mean of the coded values is 1 less than 3 times the mean of x; that is, mean $(3x-1) = 3\bar{x} - 1$.

$\mathrm{Var}(x) = \dfrac{1}{3^2} \times \mathrm{Var}(3x-1)$

$\dfrac{\Sigma x^2}{10} - 2.1^2 = \dfrac{1}{9} \times \left(\dfrac{9136}{10} - 5.3^2\right)$

$\dfrac{\Sigma x^2}{10} - 4.41 = 98.39$

$\therefore \Sigma x^2 = 1028$

We form and solve an equation knowing that
$\mathrm{Var}(x) = \dfrac{1}{a^2} \times \mathrm{Var}(ax-b)$ and $\mathrm{Var}(3x-1) = \dfrac{9136}{10} - 5.3^2$.

EXERCISE 3E

1   The range of prices of the newspapers sold at a kiosk is $0.80. After 6 p.m. all prices are reduced by 20%. Find the range of the prices after 6 p.m.

2   Find the standard deviation of x, given that $\Sigma 4x^2 = 14600$, $\Sigma 2x = 420$ and $n = 20$.

3   The values of x given in the table on the left have a standard deviation of 0.88. Find the standard deviation of the values of y.

x	2	4	6
f	a	b	c

y	7	13	19
f	a	b	c

4   The temperatures, $T°$ Celsius, at seven locations in the Central Kalahari Game Reserve were recorded at 4 p.m. one January afternoon. The values of T, correct to 1 decimal place, were:
32.1, 31.7, 31.2, 31.5, 31.9, 32.2 and 32.7.

   a   Evaluate $\Sigma 10(T - 30)$ and $\Sigma 100(T - 30)^2$.

   b   Use your answers to part **a** to calculate the standard deviation of T.

   c   By 5 p.m. the temperature at each location had dropped by exactly 0.75 °C. Find the variance of the temperatures at 5 p.m.

5   Building plots are offered for sale at $315 per square metre. The seller has to pay a lawyer's fee of $500 from the money received. Salome's plot is 240 square metres larger than Nadia's plot. How much more did the seller receive from Salome than from Nadia after paying the lawyer's fees?

6   Temperatures in degrees Celsius (°C) can be converted to temperatures in degrees Fahrenheit (°F) using the formula $F = 1.8C + 32$.

   a   The temperatures yesterday had a range of 15 °C. Express this range in degrees Fahrenheit.

   b   Temperatures elsewhere were recorded at hourly intervals in degrees Fahrenheit and were found to have mean 54.5 and variance 65.61. Find the mean and standard deviation of these temperatures in degrees Celsius.

7 Ten items were selected from each of four sections at a supermarket. Details of the prices of those items, in dollars, on 1st April and on 1st June are shown in the following table.

	1st April		1st June	
	Mean	SD	Mean	SD
Bakery	2.50	0.40	2.25	0.36
Household	5.00	1.20	4.75	1.25
Tinned food	4.00	0.60	4.10	0.60
Fruit & veg	2.00	0.40	2.00	0.50

For which section's items could each of the following statements be true? Briefly explain each of your answers.

a The total cost of the items did not change.

b The price of each item changed by the same amount.

c The proportional change in the price of each item was the same.

8 The lengths of 45 ropes used at an outdoor recreational centre can be extended by 30% when stretched. The sum of the squares of their stretched lengths is 0.0507 km^2 and their natural lengths, x metres, are summarised by $\Sigma(x-20)^2 = 1200$. Find the mean natural length of these 45 ropes.

9 Over a short period of time in 2016, the value of the pound sterling (£) fell by 15.25% against the euro (€). Find the percentage change in the value of the euro against the pound over this same period.

Appendix to Section 3.3

In this appendix, we show how the two formulae for variance are equivalent. For simplicity, we will assume that $n = 3$.

If we denote our three numbers by x_1, x_2 and x_3, then $\bar{x} = \frac{1}{3}(x_1 + x_2 + x_3)$.

Variance is defined by $\text{Var}(x) = \frac{\Sigma(x-\bar{x})^2}{n}$, so if we expand the brackets and rearrange, we get

$$\text{Var}(x) = \frac{1}{3}\left[(x_1 - \bar{x})^2 + (x_2 - \bar{x})^2 + (x_3 - \bar{x})^2\right]$$

$$= \frac{1}{3}\left[x_1^2 - 2\bar{x}x_1 + \bar{x}^2 + x_2^2 - 2\bar{x}x_2 + \bar{x}^2 + x_3^2 - 2\bar{x}x_3 + \bar{x}^2\right]$$

$$= \frac{1}{3}\left[x_1^2 + x_2^2 + x_3^2 - 2\bar{x}(x_1 + x_2 + x_3) + 3\bar{x}^2\right]$$

$$= \frac{x_1^2 + x_2^2 + x_3^2}{3} - 2\bar{x}\left(\frac{x_1 + x_2 + x_3}{3}\right) + \bar{x}^2 \quad \text{Note: the term in brackets is equal to } \bar{x}.$$

$$= \frac{x_1^2 + x_2^2 + x_3^2}{3} - 2\bar{x}^2 + \bar{x}^2$$

$$= \frac{\Sigma x^2}{3} - \bar{x}^2, \text{ which is the alternative formula } \frac{\Sigma x^2}{n} - \bar{x}^2 \text{ in the case where } n = 3.$$

Try showing that the two formulae for variance are equivalent for the simple case where $n = 2$, and then challenge yourself by taking on $n = 4$ or larger. Can you generalise this argument to an arbitrary value of n?

Checklist of learning and understanding

- Commonly used measures of variation are the range, interquartile range and standard deviation.
- A box-and-whisker diagram shows the smallest and largest values, the lower and upper quartiles and the median of a set of data.
- For ungrouped data, the median Q_2 is at the $\left(\dfrac{n+1}{2}\right)$th value.
- For grouped data with total frequency $n = \Sigma f$, the quartiles are at the following values.
  - Lower quartile Q_1 is at $\dfrac{n}{4}$ or $\dfrac{1}{4}\Sigma f$.
  - Middle quartile Q_2 is at $\dfrac{n}{2}$ or $\dfrac{1}{2}\Sigma f$.
  - Upper quartile Q_3 is at $\dfrac{3n}{4}$ or $\dfrac{3}{4}\Sigma f$.
  - IQR $= Q_3 - Q_1$
- For ungrouped data:
  Standard deviation $= \sqrt{\text{Variance}} = \sqrt{\dfrac{\Sigma(x-\bar{x})^2}{n}} = \sqrt{\dfrac{\Sigma x^2}{n} - \bar{x}^2}$, where $\bar{x} = \dfrac{\Sigma x}{n}$.
- For grouped data:
  Standard deviation $= \sqrt{\text{Variance}} = \sqrt{\dfrac{\Sigma(x-\bar{x})^2 f}{\Sigma f}} = \sqrt{\dfrac{\Sigma x^2 f}{\Sigma f} - \bar{x}^2}$, where $\bar{x} = \dfrac{\Sigma xf}{\Sigma f}$.
- For datasets x and y with n_x and n_y values, respectively:
  Mean $= \dfrac{\Sigma x + \Sigma y}{n_x + n_y}$ and Variance $= \dfrac{\Sigma x^2 + \Sigma y^2}{n_x + n_y} - \left(\dfrac{\Sigma x + \Sigma y}{n_x + n_y}\right)^2$.
- The formulae for ungrouped and grouped coded data can be summarised by:
  $$\text{Var}(x) = \text{Var}(x-b) \text{ and}$$
  $$\text{Var}(x) = \dfrac{1}{a^2} \times \text{Var}(ax-b) \text{ or } \text{Var}(ax-b) = a^2 \times \text{Var}(x)$$
- For ungrouped coded data:
  $$\dfrac{\Sigma x^2}{n} - \bar{x}^2 = \dfrac{\Sigma(x-b)^2}{n} - \left(\dfrac{\Sigma(x-b)}{n}\right)^2 \text{ and}$$
  $$\dfrac{\Sigma x^2}{n} - \bar{x}^2 = \dfrac{1}{a^2}\left[\dfrac{\Sigma(ax-b)^2}{n} - \left(\dfrac{\Sigma(ax-b)}{n}\right)^2\right]$$
- For grouped coded data:
  $$\dfrac{\Sigma x^2 f}{\Sigma f} - \bar{x}^2 = \dfrac{\Sigma(x-b)^2 f}{\Sigma f} - \left(\dfrac{\Sigma(x-b)f}{\Sigma f}\right)^2 \text{ and}$$
  $$\dfrac{\Sigma x^2 f}{\Sigma f} - \bar{x}^2 = \dfrac{1}{a^2}\left[\dfrac{\Sigma(ax-b)^2 f}{\Sigma f} - \left(\dfrac{\Sigma(ax-b)f}{\Sigma f}\right)^2\right]$$

Chapter 3: Measures of variation

END-OF-CHAPTER REVIEW EXERCISE 3

1. Three boys and seven girls are asked how much money they have in their pockets. The boys have $2.50 each and the mean amount that the 10 children have is $3.90.

   a. Show that the girls have a total of $31.50. [1]

   b. Given that the seven girls have equal amounts of money, find the standard deviation of the amounts that the 10 children have. [3]

2. Jean Luc was asked to record the times of 20 athletes in a long distance race. He started his stopwatch when the race began and then went to sit in the shade, where he fell asleep. On waking, he found that x athletes had already completed the race but he was able to record the times taken by all the others.

   State the possible value(s) of x if Jean Luc was able to use his data to calculate:

   a. the variance of the times taken by the 20 athletes [1]

   b. the interquartile range of the times taken by the 20 athletes. [2]

3. The quiz marks of nine students are written down in ascending order and it is found that the range and interquartile range are equal. Find the greatest possible number of distinct marks that were obtained by the nine students. [2]

4. Two days before a skiing competition, the depths of snow, x metres, at 32 points on the course were measured and it was discovered that the numerical values of Σx and Σx^2 were equal.

   a. Given that the mean depth of snow was 0.885 m, find the standard deviation of x. [2]

   b. Snow fell the day before the competition, increasing the depth over the whole course by 1.5 cm. Explain what effect this had on the mean and on the standard deviation of x. [2]

5. The following box plots summarise the percentage scores of a class of students in the three Mathematics tests they took this term.

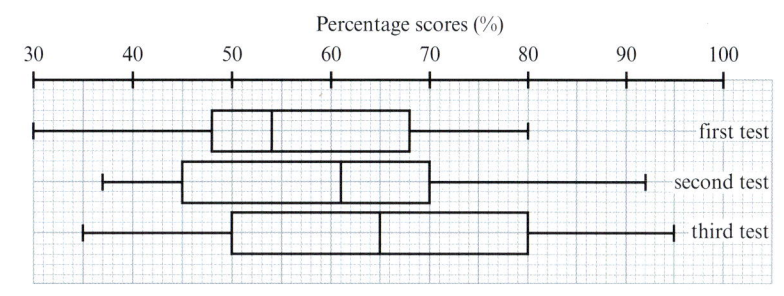

   a. Describe the progress made by the class in Mathematics tests this term. [2]

   b. Which of the tests has produced the least skewed set of scores? [1]

   c. What type of skew do the scores in each of the other two tests have? [2]

6. The following table shows the mean and standard deviation of the lengths of 75 adult puff adders (*Bitis arietans*), which are found in Africa and on the Arabian peninsula.

	Frequency	Mean (cm)	SD (cm)
African	60	102.7	6.8
Arabian	15	78.8	4.2

   a. Find the mean length of the 75 puff adders. [3]

   b. The lengths of individual African puff adders are denoted by x_f and the lengths of individual Arabian puff adders by x_b. By first finding Σx_f^2 and Σx_b^2, calculate the standard deviation of the lengths of all 75 puff adders. [5]

7 The scores obtained by 11 people throwing three darts each at a dartboard are 54, 46, 43, 52, 180, 50, 41, 56, 52, 49 and 54.

   a Find the range, the interquartile range and the standard deviation of these scores. [4]

   b Which measure in part **a** best summarises the variation of the scores? Explain why you have chosen this particular measure. [2]

8 The heights, x cm, of a group of 28 people were measured. The mean height was found to be 172.6 cm and the standard deviation was found to be 4.58 cm. A person whose height was 161.8 cm left the group.

   i Find the mean height of the remaining group of 27 people. [2]

   ii Find Σx^2 for the original group of 28 people. Hence find the standard deviation of the heights of the remaining group of 27 people. [4]

   Cambridge International AS & A Level Mathematics 9709 Paper 63 Q4 June 2014

9 120 people were asked to read an article in a newspaper. The times taken, to the nearest second, by the people to read the article are summarised in the following table.

Time (seconds)	1–25	26–35	36–45	46–55	56–90
Number of people	4	24	38	34	20

Calculate estimates of the mean and standard deviation of the reading times. [5]

Cambridge International AS & A Level Mathematics 9709 Paper 62 Q2 June 2015

10 The weights, in kilograms, of the 15 basketball players in each of two squads, A and B, are shown below.

Squad A	97	98	104	84	100	109	115	99	122	82	116	96	84	107	91
Squad B	75	79	94	101	96	77	111	108	83	84	86	115	82	113	95

   i Represent the data by drawing a back-to-back stem-and-leaf diagram with squad A on the left-hand side of the diagram and squad B on the right-hand side. [4]

   ii Find the interquartile range of the weights of the players in squad A. [2]

   iii A new player joins squad B. The mean weight of the 16 players in squad B is now 93.9 kg. Find the weight of the new player. [3]

   Cambridge International AS & A Level Mathematics 9709 Paper 62 Q5 November 2015 [Adapted]

11 The heights, x cm, of a group of 82 children are summarised as follows.

   $\Sigma(x-130) = -287$, standard deviation of $x = 6.9$.

   i Find the mean height. [2]

   ii Find $\Sigma(x-130)^2$. [2]

   Cambridge International AS & A Level Mathematics 9709 Paper 63 Q2 June 2010

12 A sample of 36 data values, x, gave $\Sigma(x-45) = -148$ and $\Sigma(x-45)^2 = 3089$.

   i Find the mean and standard deviation of the 36 values. [3]

   ii One extra data value of 29 was added to the sample. Find the standard deviation of all 37 values. [4]

   Cambridge International AS & A Level Mathematics 9709 Paper 62 Q3 June 2011

13 The ages, x years, of 150 cars are summarised by $\Sigma x = 645$ and $\Sigma x^2 = 8287.5$. Find $\Sigma(x - \bar{x})^2$, where $\bar{x}$ denotes the mean of x. [4]

Cambridge International AS & A Level Mathematics 9709 Paper 62 Q1 June 2012

14 A set of data values is 152, 164, 177, 191, 207, 250 and 258.

Compare the proportional change in the standard deviation with the proportional change in the interquartile range when the value 250 in the data set is increased by 40%. [5]

15 A shop has in its stock 80 rectangular celebrity posters. All of these posters have a width to height ratio of $1:\sqrt{2}$, and their mean perimeter is 231.8 cm.
Given that the sum of the squares of the widths is 200120 cm², find the standard deviation of the widths of the posters. [4]

16 At a village fair, visitors were asked to guess how many sweets are in a glass jar. The best six guesses were 180, 211, 230, 199, 214 and 166.

a Show that the mean of these guesses is 200, and use $SD = \sqrt{\dfrac{\Sigma(x-\bar{x})^2}{n}}$ to calculate the standard deviation. [4]

b The jar actually contained 202 sweets. Without further calculation, write down the mean and the standard deviation of the errors made by these six visitors. Explain why no further calculations are required to do this. [4]

17 The number of women in senior management positions at a number of companies was investigated. The number of women at each of the 25 service companies and at each of the 16 industrial companies are denoted by w_S and w_I, respectively. The findings are summarised by the totals:
$\Sigma(w_S - 5)^2 = 28$, $\Sigma(w_S - 5) = 15$, $\Sigma(w_I - 3)^2 = 12$ and $\Sigma(w_I - 3) = -4$.

a Show that there are, on average, more than twice as many women in senior management positions at the service companies than at the industrial companies. [3]

b Show that $\Sigma w_S^2 \neq (\Sigma w_S)^2$ and that $\Sigma w_I^2 \neq (\Sigma w_I)^2$. [5]

c Find the standard deviation of the number of women in senior management positions at all of these service and industrial companies together. [3]

18 The ages, a years, of the five members of the boy-band AlphaArise are such that $\Sigma(a - 21)^2 = 11.46$ and $\Sigma(a - 21) = -6$.

The ages, b years, of the seven members of the boy-band BetaBeat are such that $\Sigma(b - 18)^2 = 10.12$ and $\Sigma(b - 18) = 0$.

a Show that the difference between the mean ages of the boys in the two bands is 1.8 years. [3]

b Find the variance of the ages of the 12 members of these two bands. [7]

Cambridge International AS & A Level Mathematics: Probability & Statistics 1

CROSS-TOPIC REVIEW EXERCISE 1

1 Two players, A and B, both played seven matches to reach the final of a tennis tournament. The number of games that each of them won in these matches are given in the following back-to-back stem-and-leaf diagram.

```
Player A  |   | Player B    Key: 5|2|6
          | 1 | 8 9         represents 25
   2 2 0  | 2 | 1 1 3 4     games for A
   8 7 5  | 2 | 6           and 26 games
        3 | 3 |             for B
```

   a How many fewer games did player B win than player A? [1]

   b Find the median number of games won by each player. [2]

   c In a single stem-and-leaf diagram, show the number of games won by these two players in all of the 14 matches they played to reach the final. [3]

2 A total of 112 candidates took a multiple-choice test that had 40 questions. The numbers of correct answers given by the candidates are shown in the following table.

No. correct answers	0–9	10–15	16–25	26–30	31–39	40
No. candidates	18	24	27	23	19	1

   a State which class contains the lower quartile and which class contains the upper quartile. Hence, find the least possible value of the interquartile range. [3]

   b Copy and complete the following table, which shows the numbers of incorrect answers given by the candidates in the test.

No. incorrect answers	0	1–9				
No. candidates	1					

   [3]

   c Calculate an estimate of the mean number of incorrectly answered questions. [3]

3 At a factory, 50-metre lengths of cotton thread are wound onto bobbins. Due to fraying, it is common for a length, l cm, of cotton to be removed after it has been wound onto a bobbin. The following table summarises the lengths of cotton thread removed from 200 bobbins.

Length removed (l cm)	$0 \leq l < 2.5$	$2.5 \leq l < 5.0$	$5 \leq l < 10$
No. bobbins	137	49	14

   a Calculate an estimate of the mean length of cotton removed. [3]

   b Use your answer to part **a** to calculate, in metres, an estimate of the standard deviation of the length of cotton remaining on the 200 bobbins. [4]

4 People applying to a Computing college are given an aptitude test. Those who are accepted take a progress test 3 months after the course has begun. The following table gives the aptitude test scores, x, and the progress test scores, y, for a random sample of eight students, A to H.

	A	B	C	D	E	F	G	H
x	61	80	74	60	83	92	71	67
y	53	77	61	70	81	54	63	85

   a Find the interquartile range of these aptitude test scores. [1]

   b Use the summary totals $\Sigma x = 588$, $\Sigma x^2 = 44080$, $\Sigma y = 544$ and $\Sigma y^2 = 38030$ to calculate the variance of the aptitude and progress test scores when they are considered together. [3]

c The mean progress score for all the students at the college is 70 and the variance is 112. Any student who scores less than 1.5 standard deviations below the mean is sent a letter advising that improvement is needed. Which of the students A to H should the letter be sent to? [1]

5 The growth of 200 tomato plants, half of which were treated with a growth hormone, was monitored over a 5-day period and is summarised in the following graphs.

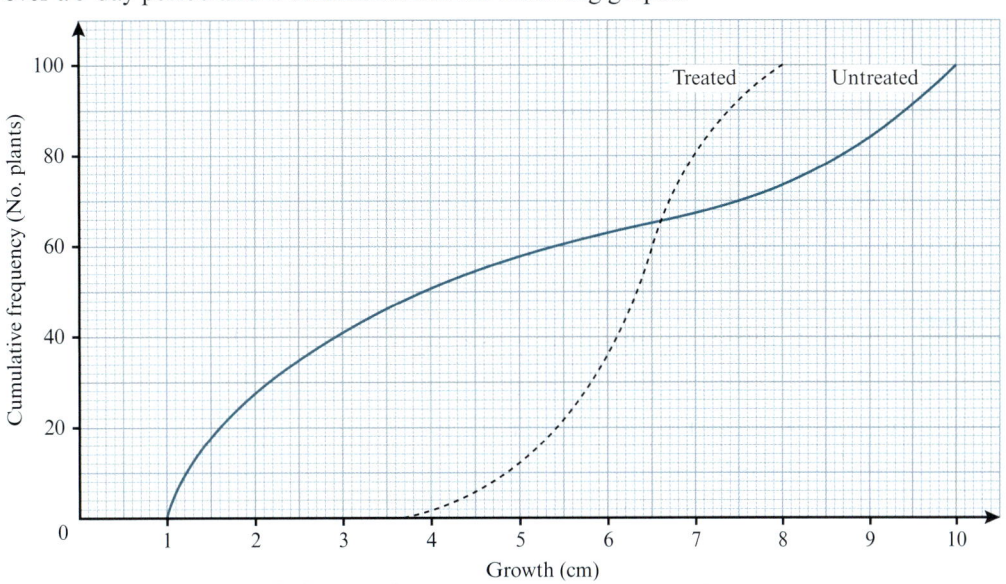

Use the graphs to describe two advantages of treating these tomato plants with the growth hormone. [2]

6 A survey of a random sample of 23 people recorded the number of unwanted emails they received in a particular week. The results are given below.

9 18 13 18 21 17 22 27 8 11 26 26 32 17 31 20 36 15 13 25 35 29 14

a Represent the data in a stem-and-leaf diagram. [3]

b Draw, on graph paper, a box-and-whisker diagram to represent the data. [4]

7 The volumes of water, $x \times 10^6$ litres, needed to fill six Olympic-sized pools are 2.82, 2.50, 2.75, 3.14, 3.66 and 3.07.

a Find the value of $\Sigma(x-2)$ and of $\Sigma(x-2)^2$. [2]

b Use your answers to part **a** to find the mean and the standard deviation of the volumes of water, giving both answers correct to the nearest litre. [5]

8 The speeds of 72 coaches at a certain point on their journeys between two cities were recorded. The results are given in the following table.

Speed (km/h)	⩽ 50	⩽ 54	⩽ 70	⩽ 75	⩽ 85
Cumulative frequency	0	9	41	54	72

a State the number of coaches whose speeds were between 54 and 70 km/h. [1]

b A student has illustrated the data in a cumulative frequency polygon. Find the two speeds between which the polygon has the greatest gradient. [1]

c Calculate an estimate of the lower boundary of the speeds of the fastest 25 coaches. [3]

9 The following table shows the masses, m grams, of 100 unsealed bags of plain potato crisps.

Mass (m grams)	$34.6 \leqslant m < 35.4$	$35.4 \leqslant m < 36.2$	$36.2 \leqslant m < 37.2$
No. bags (f)	20	30	50

   a Show that the heights of the columns in a histogram illustrating these data must be in the ratio $2:3:4$. [2]

   b Calculate estimates of the mean and of the standard deviation of the masses. [4]

   c Before each bag is sealed, 0.05 grams of salt is added. Find the variance of the masses of the sealed bags of salted potato crisps. [1]

10 The masses, in carats, of a sample of 200 pearls are summarised in the following cumulative frequency graph. One carat is equivalent to 200 milligrams.

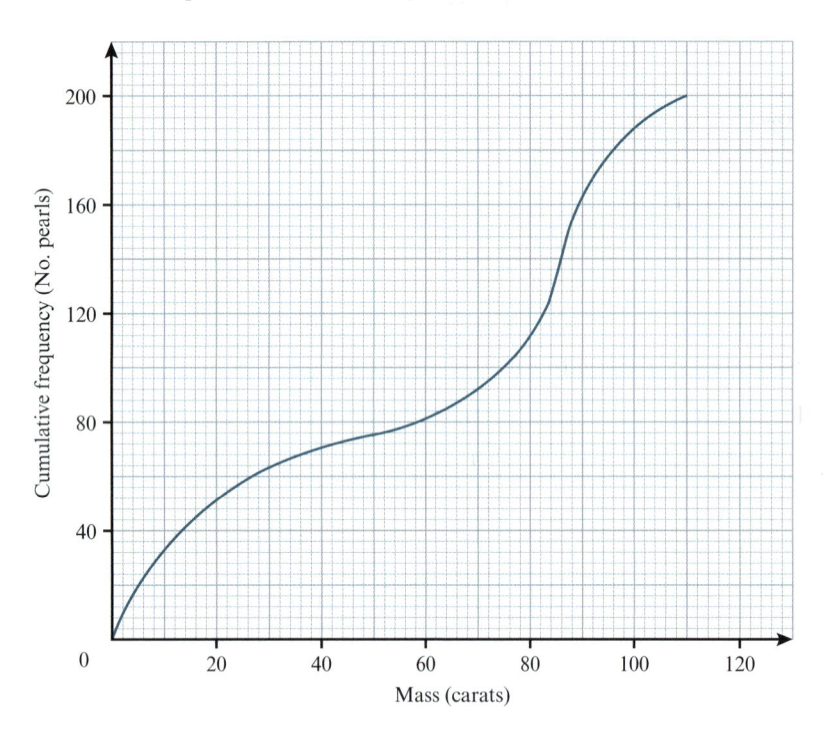

   a Use the graph to estimate, in carats:

      i   the median mass of the pearls [1]

      ii the interquartile range of the masses. [2]

   b To qualify as a 'paragon', a pearl must be flawless and weigh at least 20 grams. Use the graph to estimate the largest possible number of paragons in the sample. [2]

11 The amounts spent, S dollars, by six customers at a hairdressing salon yesterday were as follows.

   12.50, 15.75, 41.30, 34.20, 10.80, 40.85.

   Each of the customers paid with a $50 note and each received the correct change, which is denoted by C.

   a Find, in dollars, the value of $\bar{S} + \bar{C}$ and of $\bar{S} - \bar{C}$. [3]

   b Explain why the standard deviation of S and the standard deviation of C are identical. [3]

12 The numbers of goals, x, scored by a team in each of its previous 25 games are summarised by the totals $\Sigma(x-1)^2 = 30$ and $\Sigma(x-1) = 12$.

   a Find the mean number of goals that the team scored per game. [2]

   b Find the value of Σx^2. [3]

   c Find the value of a and of b in the following table, which shows the frequencies of the numbers of goals scored by the team.

No. goals	0	1	2	3	>3
No. games (f)	5	a	b	4	0

[2]

13 The lengths of some insects of the same type from two countries, X and Y, were measured. The stem-and-leaf diagram shows the results.

```
                   Country X              Country Y
(10)            9 7 6 6 6 4 4 4 3 2 | 80 |
(18)      8 8 8 7 7 6 6 5 5 5 4 4 3 3 3 2 2 0 | 81 | 1 1 2 2 3 3 3 5 5 6 7 8 9        (13)
(16)        9 9 9 8 8 7 7 6 5 5 3 2 2 2 0 0 | 82 | 0 0 1 2 3 3 3 q 4 5 6 6 7 8 8      (15)
(16)        8 7 6 5 5 5 3 3 2 2 2 1 1 1 0 0 | 83 | 0 1 2 2 4 4 4 4 5 5 6 6 7 7 7 8 9  (17)
(11)              8 7 6 5 5 4 4 3 3 1 1 | 84 | 0 0 1 2 4 4 5 5 6 6 7 7 7 8 9          (15)
                                        | 85 | 1 2 r 3 3 5 5 6 6 7 8 8               (12)
                                        | 86 | 0 1 2 2 3 5 5 5 8 9 9                 (11)
```

Key: 5|81|3 means an insect from country X has length 0.815 cm
and an insect from country Y has length 0.813 cm.

   i Find the median and interquartile range of the lengths of the insects from country X. [2]

   ii The interquartile range of the lengths of the insects from country Y is 0.028 cm. Find the values of q and r. [2]

   iii Represent the data by means of a pair of box-and-whisker plots in a single diagram on graph paper. [4]

   iv Compare the lengths of the insects from the two countries. [2]

Cambridge International AS & A Level Mathematics 9709 Paper 63 Q6 June 2010

Chapter 4
Probability

In this chapter you will learn how to:

- evaluate probabilities by means of enumeration of equiprobable (i.e. equally likely) elementary events
- use addition and multiplication of probabilities appropriately
- use the terms mutually exclusive and independent events
- determine whether two events are independent
- calculate and use conditional probabilities.

Chapter 4: Probability

> **PREREQUISITE KNOWLEDGE**
>
> | Where it comes from | What you should be able to do | Check your skills |
> |---|---|---|
> | IGCSE / O Level Mathematics | Calculate the probability of a single event as either a fraction, decimal or percentage. | 1 How many 6s are expected when an ordinary fair die is rolled 180 times? |
> | | Understand and use the probability scale from 0 to 1. | |
> | | Understand relative frequency as an estimate of probability. | 2 Find the probability of obtaining a total of 4 when the scores on two ordinary fair dice are added together. |
> | | Calculate the probability of simple combined events, using possibility diagrams and tree diagrams where appropriate. | |
> | | Use language, notation and Venn diagrams to describe sets and represent relationships between sets. | 3 It is given that $A=\{1, 2, 5\}$, $B'=\{2, 4, 5\}$ and $A'=\{3, 4\}$. Using a Venn diagram, or otherwise, find $n(A \cup B')$ and $n(A' \cap B)$. |

If we do this, how likely is that?

Probability measures the likelihood of an event occurring on a scale from 0 (i.e. impossible) to 1 (i.e. certain). We write this as P(name of event), and its value can be expressed as a fraction, decimal or percentage. The greater the probability, the more likely the event is to occur.

Although we do not often calculate **probabilities** in our daily lives, we frequently assess and compare them, and this affects our behaviour. Do we have a better chance of performing well in an exam after a good night's sleep or after revising late into the night? Should you visit a doctor or is your sore throat likely to heal by itself soon?

Insurance is based on risk, which in turn is based on the probability of certain events occurring. Government spending is largely determined by the probable benefits it will bring to society.

4.1 Experiments, events and outcomes

The result of an experiment is called an outcome or **elementary event**, and a combination of these is known simply as an event.

Rolling an ordinary fair die is an experiment that has six possible outcomes: 1, 2, 3, 4, 5 or 6.

Obtaining an odd number with the die is an event that has three **favourable** outcomes: 1, 3 or 5.

Random selection and equiprobable events

The purpose of selecting objects at **random** is to ensure that each has the same chance of being selected. This method of selection is called **fair** or **unbiased**, and the selection of any particular object is said to be equally likely or **equiprobable**.

> **KEY POINT 4.1**
>
> When one object is randomly selected from n objects, P(selecting any particular object) $= \dfrac{1}{n}$.

The probability that an event occurs is equal to the proportion of equally likely outcomes that are favourable to the event.

> **KEY POINT 4.2**
>
> $$P(\text{event}) = \frac{\text{Number of favourable equally likely outcomes}}{\text{Total number of equally likely outcomes}}$$

Consider randomly selecting 1 student from a group of 19, where 11 are boys and eight are girls.

There are 19 possible outcomes: 11 are favourable to the event *selecting a boy* and eight are favourable to the event *selecting a girl*, as shown in the following table.

Event/outcome	Probability	Description
Selecting any particular boy	$\frac{1}{19}$	These three outcomes are equally likely.
Selecting any particular girl	$\frac{1}{19}$	
Selecting any particular student	$\frac{1}{19}$	
Selecting a boy	$\frac{11}{19}$	11 of the 19 equally likely outcomes are favourable to this event.
Selecting a girl	$\frac{8}{19}$	8 of the 19 equally likely outcomes are favourable to this event.

> **TIP**
>
> The word *particular* specifies one object. It does not matter whether that object is a boy, a girl or a student; each has a $\frac{1}{19}$ chance of being selected.

Exhaustive events

A set of events that contains all the possible outcomes of an experiment is said to be exhaustive. In the special case of event A and its **complement**, not A, the sum of their probabilities is 1 because one of them is certain to occur. Recall that the notation used for the complement of set A is A'.

Examples of complementary exhaustive events are shown in the following table.

> **KEY POINT 4.3**
>
> $P(A) + P(\text{not } A) = 1$
> or
> $P(A) + P(A') = 1$

Experiment	Exhaustive events A	Exhaustive events A'	Probabilities
Toss a fair coin	heads	tails	$\frac{1}{2} + \frac{1}{2} = 1$
Roll a fair die	less than 2	2 or more	$\frac{1}{6} + \frac{5}{6} = 1$
Play a game of chess	win	not win	$P(\text{win}) + P(\text{not win}) = 1$

Trials and expectation

Each repeat of an experiment is called a **trial**. The proportion of trials in which an event occurs is its **relative frequency**, and we can use this as an estimate of the probability that the event occurs.

If we know the probability of an event occurring, we can estimate the number of times it is likely to occur in a series of trials. This is a statement of our **expectation**.

> **KEY POINT 4.4**
>
> In n trials, event A is expected to occur $n \times P(A)$ times.

WORKED EXAMPLE 4.1

The probability of rain on any particular day in a mountain village is 0.2.

On how many days is rain not expected in a year of 365 days?

Answer

$n = 365$ and P(does not rain) $= 1 - 0.2 = 0.8$

$365 \times 0.8 = 292$ days .... We multiply the probability of the event by the number of days in a year.

EXPLORE 4.1

We can see how closely expectation matches with what happens in practice by conducting simple experiments using a fair coin (or an ordinary fair die).

Toss the coin 10 times and note as a decimal the proportion of heads obtained. Repeat this and note the proportion of heads obtained in 20 trials. Continue doing this so that you have a series of decimals for the proportion of heads obtained in 10, 20, 30, 40, 50, … trials.

Represent these proportions on a graph by plotting them against the total number of trials conducted. How do your results compare with the expected proportion of heads?

For trials with a die, draw a graph to represent the proportions of odd numbers obtained.

EXERCISE 4A

1. A teacher randomly selects one student from a group of 12 boys and 24 girls. Find the probability that the teacher selects:

   a a particular boy   b a girl.

2. United's manager estimates that the team has a 65% chance of winning any particular game and an 85% chance of not drawing any particular game.

   a What are the manager's estimates most likely to be based on?

   b If the team plays 40 games this season, find the manager's expectation of the number of games the team will lose.

   c If the team loses one game more than the manager expects this season, explain why this does not necessarily mean that they performed below expectation.

3. Katya randomly picks one of the 10 cards shown.

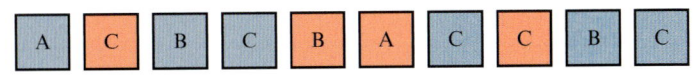

   A C B C B A C C B C

   If she repeats this 40 times, how many times is Katya expected to pick a card that is not blue and does not have a letter B on it?

4   A numbered wheel is divided into eight sectors of equal size, as shown. The wheel is spun until it stops with the arrow pointing at one of the numbers.

    Axel decides to spin the wheel 400 times.

    a   Find the number of times the arrow is not expected to point at a 4.

    b   How many more times must Axel spin the wheel so that the expected number of times that the arrow points at a 4 is at least 160?

5   A bag contains black and white counters, and the probability of selecting a black counter is $\frac{1}{6}$.

    a   What is the smallest possible number of white counters in the bag?

    b   Without replacement, three counters are taken from the bag and they are all black. What is the smallest possible number of white counters in the bag?

6   When a coin is randomly selected from a savings box, each coin has a 98% chance of not being selected. How many coins are in the savings box?

PS 7   A set of data values is 8, 13, 17, 18, 24, 32, 34 and 38. Find the probability that a randomly selected value is more than one standard deviation from the mean.

PS 8   One student is randomly selected from a school that has 837 boys. The probability that a girl is selected is $\frac{4}{7}$. Find the probability that a particular boy is selected.

> **REWIND**
>
> We studied the mean in Chapter 2, Section 2.2 and standard deviation in Chapter 3, Section 3.3.

4.2 Mutually exclusive events and the addition law

To find the probability that event A or event B occurs, we can simply add the probabilities of the two events together, but only if A and B are **mutually exclusive**.

Mutually exclusive events have no common favourable outcomes, which means that it is not possible for both events to occur, so $P(A \text{ and } B) = 0$.

For example, when we roll an ordinary die, the events 'even number = {2, 4, 6}' and 'factor of 5 = {1, 5}' are mutually exclusive because they have no common favourable outcomes. It is not possible to roll a number that is even *and* a factor of 5. We say that the intersection of these two sets is empty. Therefore:

$$P(\text{even or factor of 5}) = P(\text{even}) + P(\text{factor of 5})$$

Events are not mutually exclusive if they have at least one common favourable outcome, which means that it is possible for both events to occur, so $P(A \text{ and } B) \neq 0$.

For example, when we roll an ordinary die, the events 'odd number = {1, 3, 5}' and 'factor of 5 = {1, 5}' are not mutually exclusive because they do have common favourable outcomes. It is possible to roll a number that is odd *and* a factor of 5. We say that the intersection of these two sets is not empty. Therefore:

$$P(\text{odd or factor of 5}) \neq P(\text{odd}) + P(\text{factor of 5})$$

Chapter 4: Probability

KEY POINT 4.5

The addition law for mutually exclusive events is $P(A \text{ or } B) = P(A) + P(B)$.

This can be extended for any number of mutually exclusive events:
$P(A \text{ or } B \text{ or } C \text{ or } \ldots) = P(A) + P(B) + P(C) + \ldots$

Venn diagrams

Venn diagrams are useful tools for solving problems in probability. We can use them to show favourable outcomes or the number of favourable outcomes or the probabilities of particular events.

The number of outcomes favourable to event A is denoted by $n(A)$.

The set of outcomes that are not favourable to event A is the complement of A, denoted by A'.

The following Venn diagrams illustrate various sets and their complements.

TIP

The universal set $\mathscr{E}$ represents the complete set of outcomes and is called the *possibility space*.

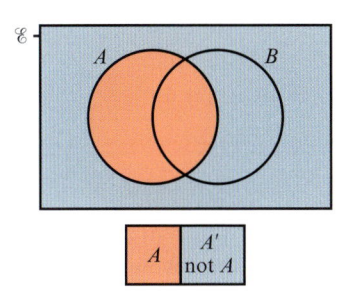

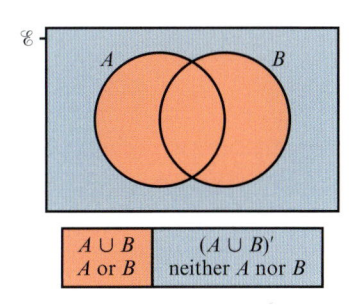

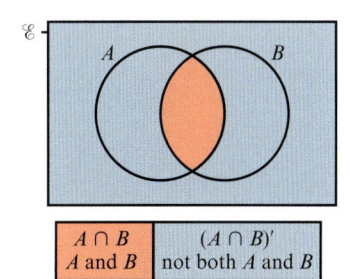

A	A'
	not A

$A \cup B$	$(A \cup B)'$
A or B	neither A nor B

$A \cap B$	$(A \cap B)'$
A and B	not both A and B

TIP

'A or B' means event A occurs or event B occurs or both occur. $A \cup B$ means 'A or B' and $A \cap B$ means 'A and B'.

KEY POINT 4.6

Using set notation, the addition law for two mutually exclusive events is $P(A \cup B) = P(A) + P(B)$.

A and B are mutually exclusive when $P(A \cap B) = 0$; that is, when $A \cap B = \varnothing$ ($\varnothing$ means the empty set).

For non-mutually exclusive events, $P(A \cup B)$ can be found by enumerating (counting) the favourable equally likely outcomes, taking care not to count any of them twice. We show how this can be done in part **b** of the following example.

WORKED EXAMPLE 4.2

One digit is randomly selected from 1, 2, 3, 4, 5, 6, 7, 8 and 9. Three possible events are:

A: a multiple of 3 is selected.

B: a factor of 8 is selected.

C: a prime number is selected.

    a Show that the only pair of mutually exclusive events from A, B and C is A and B, and find $P(A \cup B)$.

    b Find:

        i $P(A \cup C)$ **ii** $P(B \cup C)$.

Answer

$\mathcal{E} = \{1, 2, 3, 4, 5, 6, 7, 8, 9\}$

$A = \{3, 6, 9\}$, so $P(A) = \dfrac{3}{9}$.

$B = \{1, 2, 4, 8\}$, so $P(B) = \dfrac{4}{9}$.

$C = \{2, 3, 5, 7\}$, so $P(C) = \dfrac{4}{9}$.

By listing and counting the favourable outcomes, we can find the probability for each event.

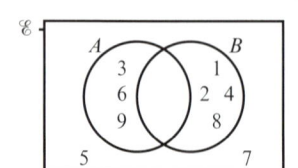

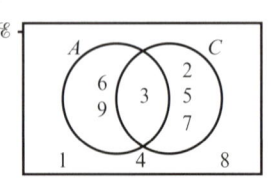

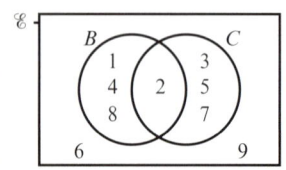

Outcomes favourable to pairs of events are shown in the three Venn diagrams opposite.

a $A \cap B = \varnothing$, so A and B are mutually exclusive.

$A \cap C \neq \varnothing$, so A and C are not mutually exclusive.

$B \cap C \neq \varnothing$, so B and C are not mutually exclusive.

Two events are mutually exclusive when they have no common favourable outcomes; that is, when their intersection is an empty set.

$P(A \cup B) = P(A \text{ or } B)$
$= P(A) + P(B)$
$= \dfrac{3}{9} + \dfrac{4}{9}$
$= \dfrac{7}{9}$

We can use the addition law because A and B are mutually exclusive events.

b Both parts of this question can be answered using the lists of elements or the previous Venn diagrams.

i $n(A \cup C) = n(A) + n(C) - n(A \cap C) = 3 + 4 - 1 = 6$

$P(A \cup C) = P(A) + P(C) - P(A \cap C) = \dfrac{3}{9} + \dfrac{4}{9} - \dfrac{1}{9} = \dfrac{6}{9}$ or $\dfrac{2}{3}$

Set A contains 3 of the 9 elements.

Set C contains 4 of the 9 elements.

Set A and set C have 1 of the 9 elements in common.

ii $n(B \cup C) = n(B) + n(C) - n(B \cap C) = 4 + 4 - 1 = 7$

$P(B \cup C) = P(B) + P(C) - P(B \cap C) = \dfrac{4}{9} + \dfrac{4}{9} - \dfrac{1}{9} = \dfrac{7}{9}$

Set B contains 4 of the 9 elements.

Set C contains 4 of the 9 elements.

Set B and set C have 1 of the 9 elements in common.

In the first part of **b** above, we subtracted $n(A \cap C)$ because the common elements in $(A \cap C)$ have been counted in $n(A)$ and in $n(C)$.

We follow the same steps when working directly with probabilities.

This also applies to events that are mutually exclusive, where the number of common elements is equal to zero.

So in general, for any two events A and B:

$n(A \cup B) = n(A) + n(B) - n(A \cap B)$ and

$P(A \cup B) = P(A) + P(B) - P(A \cap B)$.

Chapter 4: Probability

WORKED EXAMPLE 4.3

In a survey, 50% of the participants own a desktop (D), 60% own a laptop (L) and 15% own both.

What percentage of the participants owns neither a desktop nor a laptop?

Answer

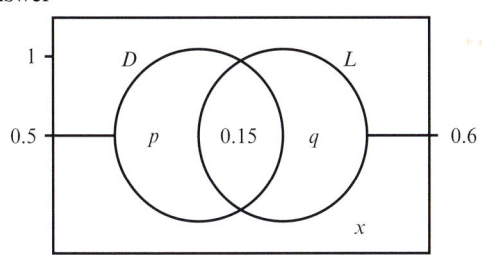

The Venn diagram shows the given information, where p, q and x represent, respectively, the percentage that own a desktop only, a laptop only and neither of these.

$p = 0.5 - 0.15 = 0.35$

$q = 0.6 - 0.15 = 0.45$

$x = 1 - (0.35 + 0.15 + 0.45) = 0.05$ or 5%

∴ 5% of the participants own neither a desktop nor a laptop.

> **TIP**
>
> The symbol ∴ means 'therefore'.

WORKED EXAMPLE 4.4

Forty children were each asked which fruits they like from apples (A), bananas (B) and cherries (C).

The following Venn diagram shows the number of children that like each type of fruit.

Find the probability that a randomly selected child likes apples or bananas.

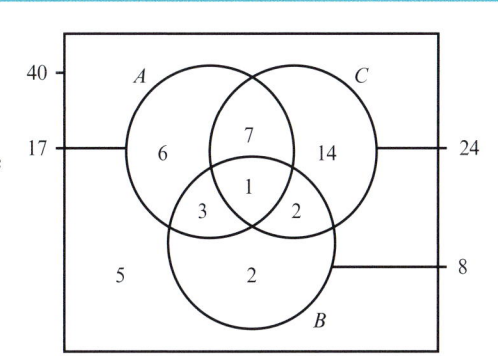

Answer

$P(A \cup B) = P(A) + P(B) - P(A \cap B)$

$= \dfrac{17}{40} + \dfrac{8}{40} - \dfrac{4}{40}$

$= \dfrac{21}{40}$

$n(A) = 17$, $n(B) = 8$ and $n(A \cap B) = 4$.

$n(A \cup B) \neq 17 + 8$ because $(A \cap B) \neq \emptyset$.

There are $17 + 8 - 4 = 21$ children who like apples or bananas.

Alternatively, we can add up the numbers in the A and B circles:
$6 + 7 + 3 + 1 + 2 + 2 = 21$.

KEY POINT 4.7

For any two events, A and B, $P(A \text{ or } B) = P(A \cup B) = P(A) + P(B) - P(A \cap B)$.

EXERCISE 4B

1 Find the probability that the number rolled with an ordinary fair die is:

   a a prime number or a 4   **b** a square number or a multiple of 3   **c** more than 3 or a factor of 8.

2 A group of 40 students took a test in Economics. The following Venn diagram shows that 19 boys (B) took the test and that seven students failed the test (F).

   a Describe the 21 students who are members of the set B'.

   b Find the probability that a randomly selected student is a boy or someone who failed the test.

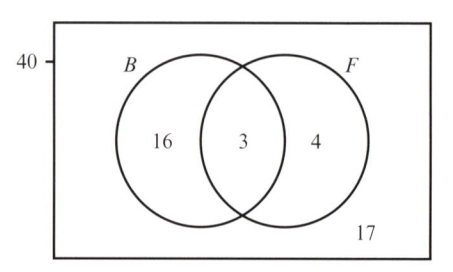

3 The following table gives information about all the animals on a farm.

	Male	Female
Goats	5	25
Sheep	3	22

   a Find the probability that a randomly selected animal is:

      i male or a goat   **ii** a sheep or female.

   b Find a different way of describing each of the two types of animal in part **a**.

4 Two ordinary fair dice are rolled and three events are:

X: the sum of the two numbers rolled is 6.

Y: the difference between the two numbers rolled is zero.

Z: both of the numbers rolled are even.

   a List the outcomes that are favourable to:

      i X and Y **ii** X and Z **iii** Y and Z.

   b What do your answers to part **a** tell you about the events X, Y and Z?

5 The letters A, B, B, B, C, D, D and E are written onto eight cards and placed in a bag. Find the probability that the letter on a randomly selected card is:

   a a vowel or in the word DOMAIN

   b a consonant or in the word DOUBLE.

6 In a group of 25 boys, nine are members of the chess club (C), eight are members of the debating club (D) and 10 are members of neither of these clubs. This information is shown in the Venn diagram.

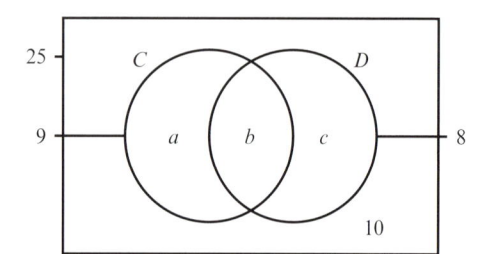

   a Find the values of a, b and c.

   b Find the probability that a randomly selected boy is:

      i a member of the chess club or the debating club

      ii a member of exactly one of these clubs.

7 Forty girls were asked to name the capital of Cuba and of Hungary; 19 knew the capital of Cuba, 20 knew the capital of Hungary and seven knew both.

   a Draw a Venn diagram showing the number of girls who knew each of these capitals.

   b Find the probability that a randomly selected girl knew:

      i   the capital of Cuba but not of Hungary
      ii just one of these capitals.

8 In a survey on pet ownership, 36% of the participants own a cat, 20% own a hamster but not a cat, and 8% own a hamster and a cat. What percentage of the participants owns neither a hamster nor a cat?

9 A garage repaired 132 vehicles last month. The number of vehicles that required electrical (E), mechanical (M) and bodywork (B) repairs are given in the diagram opposite.

   Find the probability that a randomly selected vehicle required:

   a mechanical or bodywork repairs
   b no bodywork repairs
   c exactly two types of repair.

10 The 100 students at a technical college must study at least one subject from Pure Mathematics (P), Statistics (S) and Mechanics (M). The numbers studying these subjects are given in the diagram opposite.

    a Who does the number 17 in the diagram refer to?

    b Find the probability that a randomly selected student studies:

       i   Pure Mathematics or Mechanics
       ii exactly two of these subjects.

    c List the three subjects in ascending order of popularity.

11 Events X and Y are such that $P(X) = 0.5$, $P(Y) = 0.6$ and $P(X \cap Y) = 0.2$.

    a State, giving a reason, whether events X and Y are mutually exclusive.
    b Using a Venn diagram, or otherwise, find $P(X \cup Y)$.
    c Find the probability that X or Y, but not both, occurs.

12 A, B and C are events where $P(A) = 0.3$, $P(B) = 0.4$, $P(C) = 0.3$, $P(A \cap B) = 0.12$, $P(A \cap C) = 0$ and $P(B \cap C) = 0.1$.

    a State which pair of events from A, B and C is mutually exclusive.
    b Using a Venn diagram, or otherwise, find $P[(A \cup B \cup C)']$, which is the probability that neither A nor B nor C occurs.

13 The diagram opposite shows a 30 cm square board with two rectangular cards attached. The 15 cm by 20 cm card covers one-quarter of the 8 cm by 12 cm card.

    A dart is randomly thrown at the board, so that it sticks within its perimeter. Use areas to calculate the probability that the dart pierces:

    a both cards
    b at least one of the cards
    c exactly one of the cards.

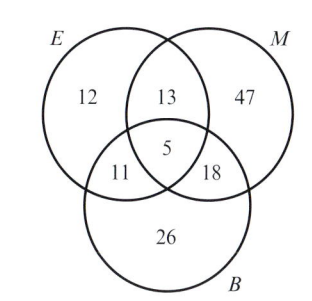

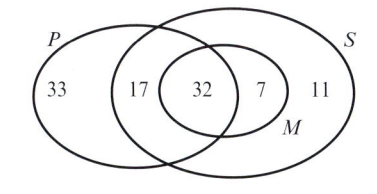

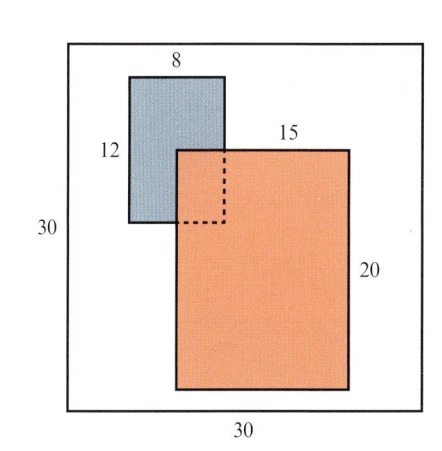

14 Given that $P(A) = 0.4$, $P(B) = 0.7$ and that $P(A \cup B) = 0.8$, find:

   a $P(A \cup B')$ b $P(A' \cap B)$

15 Each of 27 tourists was asked which of the countries Angola (A), Burundi (B) and Cameroon (C) they had visited. Of the group, 15 had visited Angola; 8 had visited Burundi; 12 had visited Cameroon; 2 had visited all three countries; and 21 had visited only one. Of those who had visited Angola, 4 had visited only one other country. Of those who had not visited Angola, 5 had visited Burundi only. All of the tourists had visited at least one of these countries.

   a Draw a fully labelled Venn diagram to illustrate this information.

   b Find the number of tourists in set B' and describe them.

   c Describe the tourists in set $(A \cup B) \cap C'$ and state how many there are.

   d Find the probability that a randomly selected tourist from this group had visited at least two of these three countries.

4.3 Independent events and the multiplication law

Two events are said to be **independent** if either can occur without being affected by the occurrence of the other. Examples of this are making selections with replacement and performing separate actions, such as rolling two dice.

KEY POINT 4.8

The multiplication law for independent events is $P(A \text{ and } B) = P(A \cap B) = P(A) \times P(B)$.

This can be extended for any number of independent events:

$P(A \text{ and } B \text{ and } C \text{ and } \ldots) = P(A \cap B \cap C \cap \ldots) = P(A) \times P(B) \times P(C) \times \ldots$

Consider the following bag, which contains two blue balls (B) and five white balls (W).

We will select one ball at random, replace it and then select another ball.

For the first selection: $P(B) = \frac{2}{7}$ and $P(W) = \frac{5}{7}$.

For the second selection: $P(B) = \frac{2}{7}$ and $P(W) = \frac{5}{7}$.

The tree diagram below shows how we can use the multiplication law to find probabilities.

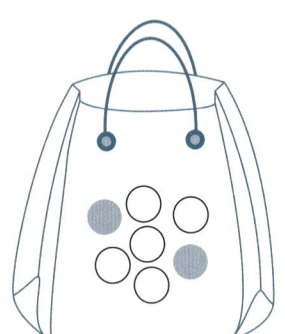

TIP

The first and second selections are made from the same seven balls, so probabilities are identical and independent.

TIP

We can denote the event '2 blue balls are selected' by BB; B and B; B & B or B, B.

$P(BB) = \frac{2}{7} \times \frac{2}{7} = \frac{4}{49}$

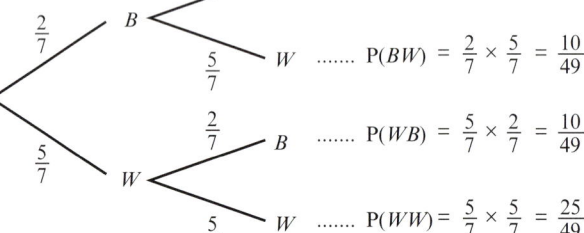

$P(BW) = \frac{2}{7} \times \frac{5}{7} = \frac{10}{49}$

$P(WB) = \frac{5}{7} \times \frac{2}{7} = \frac{10}{49}$

$P(WW) = \frac{5}{7} \times \frac{5}{7} = \frac{25}{49}$

TIP

Events BB, BW, WB and WW are exhaustive, so their probabilities sum to 1.

Multiplication of independent events is performed from left to right along the branches.

Addition of mutually exclusive events is performed vertically.

As examples:

P(different colours) = P(BW or WB) = P(BW) + P(WB) = $\left(\frac{2}{7} \times \frac{5}{7}\right) + \left(\frac{5}{7} \times \frac{2}{7}\right) = \frac{10}{49} + \frac{10}{49} = \frac{20}{49}$

P(same colours) = P(BB or WW) = P(BB) + P(WW) = $\left(\frac{2}{7} \times \frac{2}{7}\right) + \left(\frac{5}{7} \times \frac{5}{7}\right) = \frac{4}{49} + \frac{25}{49} = \frac{29}{49}$

As an alternative to using a tree diagram, we can use a possibility diagram (or *outcome space*), as shown below. The diagram shows the $7 \times 7 = 49$ equally likely outcomes and the four mutually exclusive combined events BB, BW, WB and WW.

	W	BW	BW	WW	WW	WW	WW	WW
	W	BW	BW	WW	WW	WW	WW	WW
2nd selection	W	BW	BW	WW	WW	WW	WW	WW
	W	BW	BW	WW	WW	WW	WW	WW
	W	BW	BW	WW	WW	WW	WW	WW
	B	BB	BB	WB	WB	WB	WB	WB
	B	BB	BB	WB	WB	WB	WB	WB
		B	B	W	W	W	W	W

1st selection

> **TIP**
> If we just used a 2 by 2 diagram, with B and W as the outcomes of each selection, we could not just count cells to find probabilities, because the events in those four cells would not be equally likely.

To find probabilities for combined events, we count how many of the 49 outcomes are favourable.

For example: P(at least 1 blue) = P(BB) + P(BW) + P(WB) = $\frac{4}{49} + \frac{10}{49} + \frac{10}{49} = \frac{24}{29}$.

Alternatively, P(at least 1 blue) = $1 - P(WW) = 1 - \frac{25}{49} = \frac{24}{29}$.

> **TIP**
> Probabilities are equal to the relative frequencies of the favourable outcomes.

WORKED EXAMPLE 4.5

Find the probability that the sum of the scores on three rolls of an ordinary fair die is less than 5.

Answer

P(sum < 5) = P(3) + P(4) — The lowest possible sum is 3.

P(sum = 3) = $\frac{1}{216}$

P(sum = 4) = $\frac{3}{216}$

Each roll has six equiprobable outcomes, so there are $6 \times 6 \times 6 = 216$ possible outcomes, each with a probability of $\frac{1}{216}$.

There is one way to obtain a sum of 3: (1, 1, 1).

There are three ways to obtain a sum of 4: (1, 1, 2), (1, 2, 1), (2, 1, 1).

P(sum < 5) = $\frac{1}{216} + \frac{3}{216} = \frac{1}{54}$

Cambridge International AS & A Level Mathematics: Probability & Statistics 1

WORKED EXAMPLE 4.6

Abha passes through three independent sets of traffic lights when she drives to work. The probability that she has to stop at any particular set of lights is 0.2. Find the probability that Abha:

- **a** first has to stop at the second set of lights
- **b** has to stop at exactly one set of lights
- **c** has to stop at any set of lights.

Answer

a $P(XS) = 0.8 \times 0.2 = 0.16$ ⟶ We use S and X to represent 'stopping' and 'not stopping'. For each set of lights, $P(S) = 0.2$ and $P(X) = 0.8$.

b P(has to stop at exactly 1 set of lights) ⟶ The three favourable outcomes, SXX, XSX and XXS, are equally likely.
$= P(SXX) + P(XSX) + P(XXS)$
$= 3 \times (0.2 \times 0.8 \times 0.8)$
$= 0.384$

c P(has to stop) = 1 − P(does not have to stop) ⟶ The events 'has to stop' and 'does not have to stop' are complementary.
$= 1 − P(XXX)$
$= 1 − 0.8^3$
$= 0.488$

Alternatively, P(has to stop) = P(S) + P(XS) + P(XXS).

EXERCISE 4C

1. Using a tree diagram, find the probability that exactly one head is obtained when two fair coins are tossed.

2. Two ordinary fair dice are rolled. Using a possibility diagram, find the probability of obtaining:
   - **a** two 6s
   - **b** two even numbers
   - **c** two numbers whose product is 6.

3. It is known that 8% of all new FunX cars develop a mechanical fault within a year and that 15% independently develop an electrical fault within a year. Find the probability that within a year a new FunX car develops:
   - **a** both types of fault
   - **b** neither type of fault.

4. A certain horse has a 70% chance of winning any particular race. Find the probability that it wins exactly one of its next two races.

5. The probabilities that a team wins, draws or loses any particular game are 0.6, 0.1 and 0.3, respectively.
   - **a** Find the probability that the team wins at least one of its next two games.
   - **b** If 2 points are awarded for a win, 1 point for a draw and 0 points for a loss, find the probability that the team scores a total of more than 1 point in its next two games.

6. On any particular day, there is a 30% chance of snow in Slushly. Find the probability that it snows there on:
   - **a** none of the next 3 days
   - **b** exactly one of the next 3 days.

7 Fatima will enter three sporting events at the weekend.
Her chances of winning each of them are shown in the following table.

Event	Shot put	Javelin	Discus
Chance of winning	85%	40%	64%

a Assuming that the three events are independent, find the probability that Fatima wins:

i the shot put and discus   ii the shot put and discus only   iii exactly two of these events.

b What does 'the three events are independent' mean here? Give a reason why this may not be true in real life.

8 A fair six-sided spinner, P, has edges marked 0, 1, 2, 2, 3 and 4.

A fair four-sided spinner, Q, has edges marked 0, –1, –1 and –2.

Each spinner is spun once and the numbers on which they come to rest are added together to give the score, S. Find:

a $P(S = 2)$ b $P(S^2 = 1)$

9 Letters and packages can take up to 2 days to be delivered by Speedipost couriers. The following table shows the percentage of items delivered at certain times after sending.

	Same day	After 1 day	After 2 days
Letter	40%	50%	10%
Package	15%	55%	30%

a Is there any truth in the statement 'If you post 10 letters on Monday then only nine of them will be delivered before Wednesday'? Give a reason for your answer.

b Find the probability that when three letters are posted on Monday, none of them are delivered on Tuesday.

c Find the probability that when a letter and a package are posted together, the letter arrives at least 1 day before the package.

10 The following histogram represents the results of a national survey on bus departure delay times.

Two buses are selected at random. Calculate an estimate of the probability that:

a both departures were delayed by less than 4 minutes

b at least one of the buses departed more than 7 minutes late.

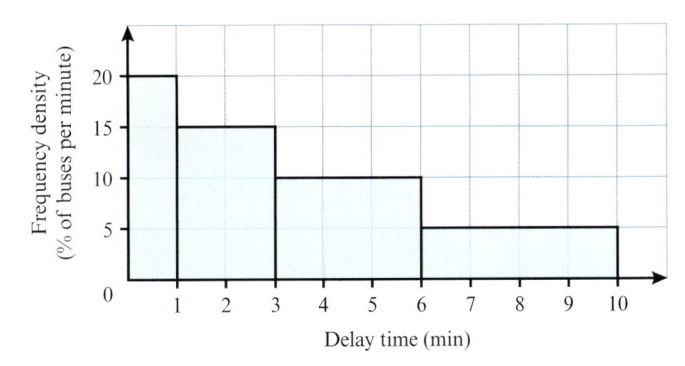

11 Praveen wants to speak on the telephone to his friend. When his friend's phone rings, he answers it with constant probability 0.6. If Praveen's friend doesn't answer his phone, Praveen will call later, but he will only try four times altogether. Find the probability that Praveen speaks with his friend:

a after making fewer than three calls   b on the telephone on this occasion.

12 Each morning, Ruma randomly selects and buys one of the four newspapers available at her local shop. Find the probability that she buys:

  a the same newspaper on two consecutive mornings

  b three different newspapers on three consecutive mornings.

13 A coin is biased such that the probability that three tosses all result in heads is $\dfrac{125}{512}$. Find the probability of obtaining no heads with three tosses of the coin.

14 In a group of five men and four women, there are three pairs of male and female business partners and three teachers, where no teacher is in a business partnership. One man and one woman are selected at random. Find the probability that they are:

  a both teachers

  b in a business partnership with each other

  c each in a business partnership but not with each other.

PS 15 A biased die in the shape of a pyramid has five faces marked 1, 2, 3, 4 and 5. The possible scores are 1, 2, 3, 4 and 5 and $P(x) = \dfrac{k-x}{25}$, where k is a constant.

  a Find, in terms of k, the probability of scoring:

  i 5   ii less than 3.

  b The die is rolled three times and the scores are added together. Evaluate k and find the probability that the sum of the three scores is less than 5.

PS 16 A game board is shown in the diagram.

Players take turns to roll an ordinary fair die, then move their counters forward from 'start' a number of squares equal to the number rolled with the die. If a player's counter ends its move on a coloured square, then it is moved back to the start.

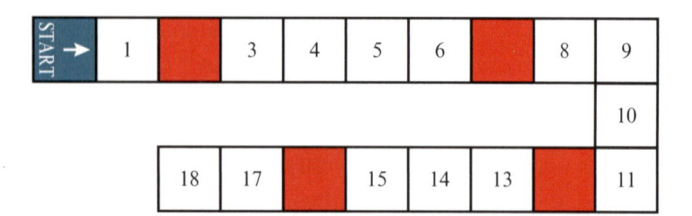

  a Find the probability that a player's counter is on 'start' after rolling the die:

  i once   ii twice.

  b Find the probability that after rolling the die three times, a player's counter is on:

  i 18   ii 17

DID YOU KNOW?

It has long been common practice to write *Q.E.D.* at the point where a mathematical proof or philosophical argument is complete. *Q.E.D.* is an initialism of the Latin phrase *quad erat demonstrandum*, meaning 'which is what had to be shown'.

Latin was used as the language of international communication, scholarship and science until well into the 18th century.

Q.E.D. does not stand for *Quite Easily Done*!

A popular modern alternative is to write W^5, an abbreviation of *Which Was What Was Wanted*.

Application of the multiplication law

The multiplication law given in Key point 4.8 can be used to show whether or not events are independent. If we can show that $P(A \cap B) = P(A) \times P(B)$, then A and B are independent, and vice versa. If, for example, $P(X) = 0.3$ and $P(Y) = 0.4$, then X and Y are independent only if $P(X \cap Y) = 0.3 \times 0.4 = 0.12$.

WORKED EXAMPLE 4.7

Events J, K and L are independent. Given that $P(J) = 0.5$, $P(K) = 0.6$ and $P(J \cap L) = 0.24$, find:

a $P(J \cap K)$ b $P(L)$ c $P(K \cap L)$.

Answer

a $P(J \cap K) = 0.5 \times 0.6$ .... We know that $P(J \cap K) = P(J) \times P(K)$.
   $= 0.3$

b $0.5 \times P(L) = 0.24$ .... We know that $P(J) \times P(L) = P(J \cap L)$.
   $P(L) = \dfrac{0.24}{0.5}$
   $= 0.48$

c $P(K \cap L) = 0.6 \times 0.48$ .... We know that $P(K \cap L) = P(K) \times P(L)$.
   $= 0.288$

Examples of independence and non-independence that you may have come across are:

- Enjoyment of sport is independent of gender if equal proportions of males and females enjoy sport.
- If unequal proportions of employed and unemployed people own cars, then car ownership is not independent of employment status – it is dependent on it.

WORKED EXAMPLE 4.8

In a group of 60 students, 27 are male (M) and 20 study History (H). The Venn diagram shows the numbers of students in these and other categories. One student is selected at random from the group. Show that the events 'a male is selected' and 'a student who studies History is selected' are independent.

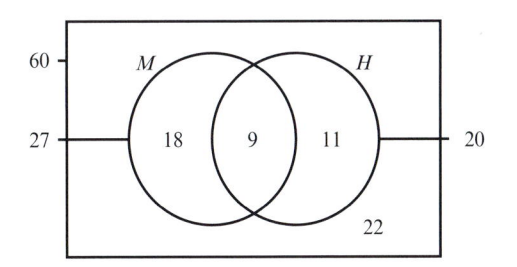

Venn diagram: 60 total; M circle contains 18 and 9; H circle contains 9 and 11; M total 27, H total 20; outside 22.

Answer

Does $P(M) \times P(H) = P(M \cap H)$? .... If the multiplication law holds for the events M and H, then they are independent.

$P(M) = \dfrac{27}{60}$, $P(H) = \dfrac{20}{60}$ and .... We state the three probabilities concerned.

$P(M \cap H) = \dfrac{9}{60}$

$P(M) \times P(H) = \dfrac{27}{60} \times \dfrac{20}{60}$ .... Then we evaluate $P(M) \times P(H)$ to show whether or not it is equal to $P(M \cap H)$.

$= \dfrac{9}{60}$

$= P(M \cap H)$

The multiplication law holds for events M and H, therefore they are independent. Q.E.D.

> **TIP**
>
> When we show that two events are independent (or not), it is important to state a conclusion in words after doing the mathematics. A short sentence like the last line of Worked example 4.8 is sufficient. Writing Q.E.D., however, is optional.

EXPLORE 4.2

In Worked example 4.8, we showed that the events M and H are independent.

For the 60 students in that particular group, we could show by similar methods whether or not the three pairs of events M and H', M' and H and M' and H' are independent.

Do this then discuss what you believe would be the most appropriate conclusion to write.

EXERCISE 4D

1. Y and Z are independent events. $P(Y) = 0.7$ and $P(Z) = 0.9$. Find $P(Y \cap Z)$.

2. Two independent events are M and N. Given that $P(M) = 0.75$ and $P(M \cap N) = 0.21$, find $P(N)$.

3. Independent events S and T are such that $P(S) = 0.4$ and $P(T') = 0.2$. Find:
   a $P(S \cap T)$ b $P(S' \cap T)$.

4. A, B and C are independent events, and it is given that $P(A \cap B) = 0.35$, $P(B \cap C) = 0.56$ and $P(A \cap C) = 0.4$.
   a Express $P(A)$ in terms of:
      i $P(B)$ ii $P(C)$.
   b Use your answers to part **a** to find:
      i $P(B)$ ii $P(A')$ iii $P(B' \cap C')$.

5. In a class of 28 children, 19 attend drama classes, 13 attend singing lessons, and six attend both drama classes and singing lessons. One child is chosen at random from the class.

   Event D is 'a child who attends drama classes is chosen'.

   Event S is 'a child who attends singing lessons is chosen'.

a Illustrate the data in an appropriate table or diagram.

b Are events D and S independent? Give a reason for your answer.

6 Each child in a group of 80 was asked whether they regularly read (R) or regularly watch a movie (M). The results are given in the Venn diagram opposite. One child is selected at random from the group. Event R is 'a child who regularly reads is selected' and event M is 'a child who regularly watches a movie is selected'.

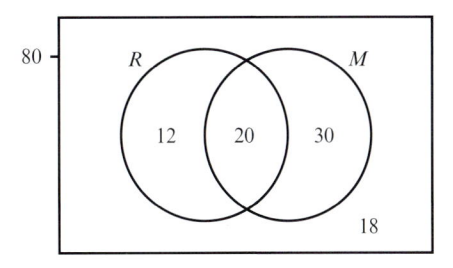

Determine, with justification, whether events R and M are independent.

7 Two fair 4-sided dice, both with faces marked 1, 2, 3 and 4, are rolled.

Event A is 'the sum of the numbers obtained is a prime number'.

Event B is 'the product of the numbers obtained is an even number'.

a Find, in simplest form, the value of P(A), of P(B) and of P($A \cap B$).

b Determine, with justification, whether events A and B are independent.

c Give a reason why events A and B are not mutually exclusive.

8 Two ordinary fair dice are rolled.

Event X is 'the product of the two numbers obtained is odd'.

Event Y is 'the sum of the two numbers obtained is a multiple of 3'.

a Determine, giving reasons for your answer, whether X and Y are independent.

b Are events X and Y mutually exclusive? Justify your answer.

9 A fair 8-sided die has faces marked 1, 2, 3, 4, 5, 6, 7 and 8. The score when the die is rolled is the number on the face that the die lands on. The die is rolled twice.

Event V is 'one of the scores is exactly 4 less than the other score'.

Event W is 'the product of the scores is less than 13'.

Determine whether events V and W are independent, justifying your answer.

10 Two hundred children are categorised by gender and by whether or not they own a bicycle. Of the 108 males, 60 own a bicycle, and altogether 90 children do not own a bicycle.

a Tabulate these data.

b Determine, giving reasons for your answer, whether ownership of a bicycle is independent of gender for these 200 children.

c What percentage of the females and what percentage of the males own bicycles?

Explain how your answers to part **c** confirm the result obtained in part **b**.

Cambridge International AS & A Level Mathematics: Probability & Statistics 1

PS 11 At an election, it was found that people voted for Party X independently of their income group. The following table shows that 12 400 people from three income groups voted altogether, and that 7440 of them voted for Party X.

	Low	Medium	High	Totals
Party X	a	b	c	7440
Totals	3100	6820	2480	12 400

Find the value of a, of b and of c.

PS 12 The speed limit at a motorway junction is 120 km/h. Information about the speeds and directions in which 207 vehicles were being driven are shown in the following table.

	North	East	South	West
Under limit	36	27	36	39
Over limit	15	15	18	21

Providing evidence to support your answer, determine which vehicles' speeds were independent of their direction of travel.

> **FAST FORWARD**
>
> We will study discrete random variables that arise from independent events in Chapter 6.

4.4 Conditional probability

The word *conditional* is used to describe a probability that is dependent on some additional information given about an outcome or event.

For example, if your friend randomly selects a letter from the word ACE, then $P(\text{selects E}) = \frac{1}{3}$.

However, if we are told that she selects a vowel, we now have a conditional probability that is not the same as P(selects E).

This conditional probability is $P(\text{selects E, given that she selects a vowel}) = \frac{1}{2}$.

Conditional probabilities are usually written using the symbol | to mean *given that*.

We read $P(A \mid B)$ as 'the probability that A occurs, given that B occurs'.

> **WORKED EXAMPLE 4.9**
>
> A child is selected at random from a group of 11 boys and nine girls, and one of the girls is called Rose. Find the probability that Rose is selected, given that a girl is selected.
>
> **Answer**
>
> $P(\text{Rose is selected} \mid \text{a girl is selected}) = \frac{1}{9}$ The additional information, 'given that a girl is selected', reduces the number of possible selections from 20 to 9, and Rose is one of those nine girls.

Chapter 4: Probability

WORKED EXAMPLE 4.10

The following table shows the numbers of students in a class who study Biology (*B*) and who study Chemistry (*C*).

	B	*B'*	Totals
C	9	8	17
C'	7	1	8
Totals	16	9	25

Represent the data in a suitable Venn diagram, and find the probability that a randomly selected student:

a studies Chemistry, given that they study Biology

b does not study Biology, given that they do not study Chemistry.

Answer

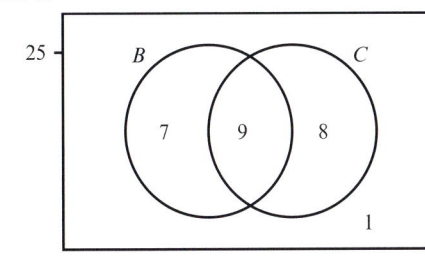

[Venn diagram: universal set = 25; B contains 7, intersection B∩C contains 9, C contains 8, outside = 1]

The number of students in each category is shown in the Venn diagram.

a $P(C \mid B) = \dfrac{9}{16}$

16 study Biology and nine of these study Chemistry.

b $P(B' \mid C') = \dfrac{1}{8}$

Eight students do not study Chemistry and one of these does not study Biology.

WORKED EXAMPLE 4.11

Two children are selected at random from a group of five boys and seven girls. Find the probability that the second child selected is a boy, given that the first child selected is:

   a a boy   b a girl.

Answer

a $P(\text{second is a boy} \mid \text{first is a boy}) = \dfrac{4}{11}$

If a boy is selected first, then the second child is selected from four boys and seven girls.

b $P(\text{second is a boy} \mid \text{first is a girl}) = \dfrac{5}{11}$

If a girl is selected first, then the second child is selected from five boys and six girls.

Cambridge International AS & A Level Mathematics: Probability & Statistics 1

EXERCISE 4E

1. One letter is randomly selected from the six letters in the word BANANA. Find the probability that:
   a an N is selected, given that an A is not selected
   b an A is selected, given that an N is not selected.

2. One hundred children were each asked whether they have brothers (B) and whether they have sisters (S). Their responses are given in the Venn diagram opposite.

   Find the probability that a randomly selected child has:

   a sisters, given that they have brothers
   b brothers, given that they do not have sisters
   c sisters or brothers, given that they do not have both.

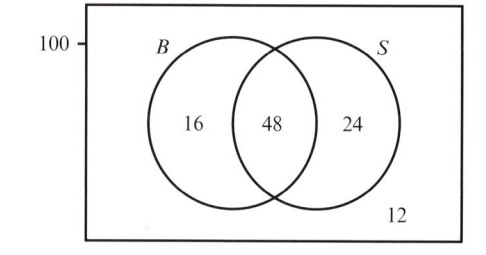

   [Venn diagram: 100 total; B only = 16, B∩S = 48, S only = 24, outside = 12]

3. Two photographs are randomly selected from a pack of 12 colour and eight black and white photographs. Find the probability that the second photograph selected is colour, given that the first is:
   a colour
   b black and white.

4. The Venn diagram opposite shows the responses of 40 girls who were asked if they have an interest in a career in nursing (N), dentistry (D) or human rights (H).

   a Find the probability that a randomly selected girl has an interest in:
      i human rights, given that she has an interest in nursing
      ii nursing, given that she has no interest in dentistry.
   b Describe any group of girls for whom dentistry is the least popular career of interest.

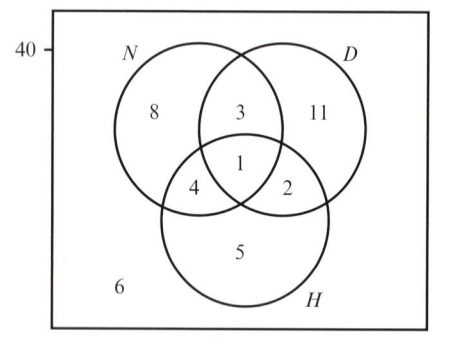

   [Venn diagram: 40 total; N only = 8, D only = 11, H only = 5, N∩D only = 3, N∩H only = 4, D∩H only = 2, N∩D∩H = 1, outside = 6]

5. The quiz marks of 40 students are represented in the following bar chart.

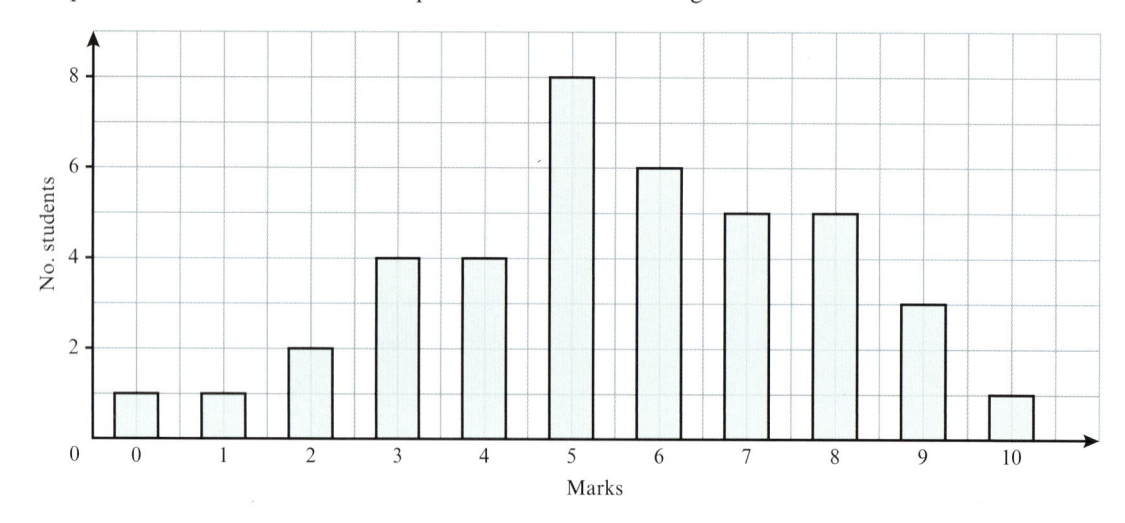

   [Bar chart: Marks vs No. students — 0:1, 1:1, 2:2, 3:4, 4:4, 5:8, 6:6, 7:5, 8:5, 9:3, 10:1]

   Two students are selected at random from the group. Find the probability that the second student:

   a scored more than 5, given that the first student did not score more than 5
   b scored more than 7, given that the first student scored more than 7.

6 The histogram shown represents the times taken, in minutes, for 115 men to complete a task.

Two men are selected at random from the group. Find the probability that the:

   a first man took less than 1 minute, given that he took less than 3 minutes

   b second man took less than 6 minutes, given that the first man took less than 1 minute.

7 At an insurance company, 60% of the staff are male (M) and 70% work full-time (FT). The following Venn diagram shows this and one other piece of information.

   a What information is given by the value 0.10 in the Venn diagram?

   b Find the value of a, of b and of c.

   c An employee is randomly selected. Find:

      i $P(M \mid FT)$ ii $P(FT \mid M')$ iii $P[(M \cap FT) \mid (M \cup FT)]$

8 Two fair triangular spinners, both with sides marked 1, 2 and 3, are spun. Given that the sum of the two numbers spun is even, find the probability that the two numbers are the same.

9 Two ordinary fair dice are rolled and the two numbers rolled are added together to give the score. Given that a player's score is greater than 6, find the probability that it is not greater than 8.

10 The circular archery target shown, on which 1, 2, 3 or 5 points can be scored, is divided into four parts of unequal area by concentric circles. The radii of the circles are 3 cm, 9 cm, 15 cm and 30 cm.

You may assume that a randomly fired arrow pierces just one of the four areas and is equally likely to pierce any part of the target.

   a Show that the probability of scoring 5 points is 0.01.

   b Find the probability of scoring 3 points, 2 points and 1 point with an arrow.

   c Given that an arrow does not score 5 points, find the probability that it scores 1 point.

   d Given that a total score of 6 points is obtained with two randomly fired arrows, find the probability that neither arrow scores 1 point.

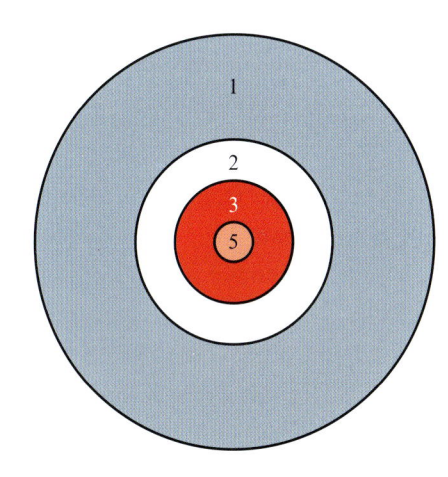

Independence and conditional probability

At the beginning of Section 4.3, 'independent' was described in quite familiar terms. In general, we can use the multiplication law given in Key point 4.8 as the definition of 'independent'. However, a more formal definition can now be given.

Events X and Y are said to be independent if each is unaffected by the occurrence of the other. If this is the case then the probability that X occurs is the same in two complementary situations:

(i) when Y occurs, and (ii) when Y does not occur.

From these, we can now say that X and Y are independent if and only if $P(X|Y) = P(X|Y')$.

Consider rolling an ordinary fair die and the events X and Y, as defined below.
X: the outcome is a square number (1 or 4).
Y: the outcome is an odd number (1, 3 or 5).
First note that 1 is odd and square, so X and Y are not mutually exclusive; but are they independent?

When Y occurs, the die shows 1, 3 or 5, so $P(X|Y) = \dfrac{1}{3}$

When Y does not occur, the die shows 2, 4 or 6, so $P(X|Y') = \dfrac{1}{3}$

$P(X) = \dfrac{1}{3}$, whether Y occurs or not.

$P(X|Y) = P(X|Y')$ means that $P(X)$ is unaffected by the occurrence of Y.
Events X and Y are not mutually exclusive, but they are independent.

EXPLORE 4.3

1. Consider rolling an ordinary fair die. In each case below, determine whether the given events are mutually exclusive, and whether they are independent.
   a 'X: a number less than 3', and 'Y: an even number'.
   b 'A: a number that is 4 or more', and 'B: a number that is not more than 4'.

2. Now consider rolling a fair 12-sided die, numbered from 1 to 12. In each case below, determine whether the given events are mutually exclusive, and whether they are independent.
   a 'X: an even number', and 'Y: a factor of 28'.
   b 'A: a prime number', and 'B: a multiple of 3'.
   c 'F: a factor of 12', and 'M: a multiple of 5'.

3. An integer from 1 and 20 inclusive is selected at random. Three events are defined as follows:
   A: the number is a multiple of 3.
   B: the number is a factor of 72.
   C: the number has exactly two digits, and at least one of those digits is a 1.
   Determine which of the three possible pairs of these events is independent.

4.5 Dependent events and conditional probability

Two events are mutually **dependent** when neither can occur without being affected by the occurrence of the other. An example of this is when we make selections without replacement; that is, when probabilities for the second selection depend on the outcome of the first selection.

The multiplication law for independent events (see Key point 4.8) is a special case of the multiplication law of probability.

The multiplication law of probability is used to find the probability that 'this and that' occurs when the events involved might not be independent.

Chapter 4: Probability

> **KEY POINT 4.9**
>
> $P(A \text{ and } B) = P(A \cap B) = P(A) \times P(B|A)$
>
> $P(B \text{ and } A) = P(B \cap A) = P(B) \times P(A|B)$
>
> $P(A) \times P(B|A) \equiv P(B) \times P(A|B)$.

WORKED EXAMPLE 4.12

Two children are randomly selected from 11 boys (B) and 14 girls (G). Find the probability that the selection consists of:

a two boys   b a boy and a girl, in any order.

Answer

a $P(2 \text{ boys}) = P(B \text{ and } B)$
    $= P(B_1) \times P(B_2 | B_1)$
    $= \dfrac{11}{25} \times \dfrac{10}{24}$
    $= \dfrac{11}{60}$

The first child is selected from 25 but the second is selected from the remaining 24. This tells us that the selections are dependent and that we must use conditional probabilities (suffices mean 1st and 2nd).

b $P(\text{a boy and a girl}) = P(B \text{ and } G) + P(G \text{ and } B)$
    $= P(B_1) \times P(G_2 | B_1) + P(G_1) \times P(B_2 | G_1)$
    $= \left(\dfrac{11}{25} \times \dfrac{14}{24}\right) + \left(\dfrac{14}{25} \times \dfrac{11}{24}\right)$
    $= \dfrac{77}{150}$

There are two different orders in which a boy and a girl can be selected, but note that $P(B \text{ and } G) = P(G \text{ and } B)$.

WORKED EXAMPLE 4.13

Every Saturday, a man invites his sister to the theatre or to the cinema. 70% of his invitations are to the theatre and 90% of these are accepted. His sister rejects 40% of his invitations to the cinema.

Find the probability that the brother's invitation is accepted on any particular Saturday.

Answer

The probability that the sister accepts depends on where she is invited to go. This tells us that our calculations must involve conditional probabilities.

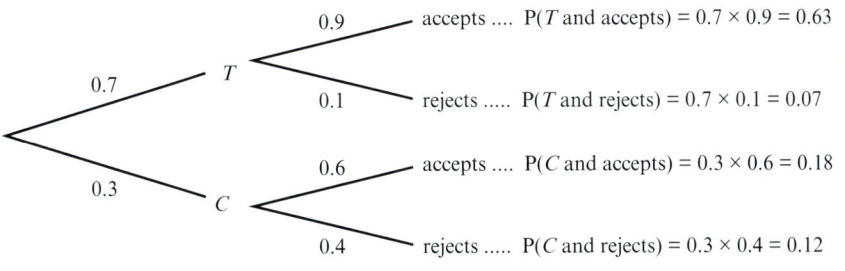

```
              0.9   accepts ....  P(T and accepts) = 0.7 × 0.9 = 0.63
       0.7  T
              0.1   rejects .....  P(T and rejects) = 0.7 × 0.1 = 0.07

              0.6   accepts ....  P(C and accepts) = 0.3 × 0.6 = 0.18
       0.3  C
              0.4   rejects .....  P(C and rejects) = 0.3 × 0.4 = 0.12
```

The given information is shown in the tree diagram, where T and C stand for 'theatre' and 'cinema'.

$P(\text{accepts}) = P(T \text{ and accepts}) + P(C \text{ and accepts})$
    $= 0.63 + 0.18$
    $= 0.81$

The sister can accept an invitation to the theatre or to the cinema.

We can use the multiplication law of probability to find conditional probabilities.

We know that $P(A \cap B) = P(A) \times P(B \mid A)$, so $P(B \mid A)$ can be found when $P(A \cap B)$ and $P(A)$ are known.

> **KEY POINT 4.10**
>
> $$P(B \mid A) = \frac{P(A \cap B)}{P(A)} \quad \text{and} \quad P(A \mid B) = \frac{P(B \cap A)}{P(B)}, \text{ where } P(A \cap B) \equiv P(B \cap A).$$

TIP

The symbol $\equiv$ means 'is identical to'.

WORKED EXAMPLE 4.14

Given that $P(A \cap B) = 0.36$ and $P(B) = 0.9$, find $P(A \mid B)$.

Answer

$$P(A \mid B) = \frac{P(B \cap A)}{P(B)}$$
$$= \frac{P(A \cap B)}{P(B)}$$
$$= \frac{0.36}{0.9}$$
$$= 0.4$$

WORKED EXAMPLE 4.15

An ordinary fair die is rolled. Find the probability that the number obtained is prime, given that it is odd.

Answer

$$P(\text{prime} \mid \text{odd}) = \frac{P(\text{odd and prime})}{P(\text{odd})}$$
$$= \frac{2}{6} \div \frac{3}{6}$$
$$= \frac{2}{3}$$

The odd numbers are 1, 3 and 5, so $P(\text{odd}) = \frac{3}{6}$. The odd prime numbers are 3 and 5, so $P(\text{odd and prime}) = \frac{2}{6}$.

TIP

Alternatively, two of the three odd numbers on a die are prime, so $P(\text{prime} \mid \text{odd}) = \frac{2}{3}$.

WORKED EXAMPLE 4.16

A boy walks to school (W) 60% of the time and cycles (C) 40% of the time. He is late to school (L), on 5% of the occasions that he walks, and he is late on 2% of the occasions that he cycles.

Given that he is late to school, find the probability that he cycles; that is, find $P(C \mid L)$.

Answer

The tree diagram shows the given information.

$P(L) = P(W \text{ and } L) + P(C \text{ and } L)$
$= 0.030 + 0.008$
$= 0.038$

The boy can arrive late by walking or by cycling.

$P(C \mid L) = \dfrac{P(C \text{ and } L)}{P(L)}$

$= \dfrac{0.008}{0.038}$

$= \dfrac{4}{19}$ or 0.211

EXERCISE 4F

1. Two ties are taken at random from a bag of three plain and five striped ties. By use of a tree diagram, or otherwise, find the probability that both ties are:

   a plain   b striped.

2. There are four toffee sweets and seven nutty sweets in a girl's pocket. Find the probability that two sweets, selected at random, one after the other, are not the same type.

3. On a library shelf there are seven novels, three dictionaries and two atlases. Two books are randomly selected without replacement from these. Find the probability that the selected books are:

   a both novels   b both dictionaries or both atlases.

4. A woman travels to work by bicycle 70% of the time and by scooter 30% of the time. If she uses her bicycle she is late 3% of the time but if she uses her scooter she is late only 2% of the time.

   a Find the probability that the woman is late for work on any particular day.

   b Given that the woman expects not to be late on approximately 223 days in a year, find the number of days in a year on which she works.

5. Two children are randomly selected from a group of five boys and seven girls. Determine which is more likely to be selected:

   a two boys or two girls?

   b the two youngest girls or the two oldest boys?

6 A boy has five different pairs of shoes mixed up under his bed. Find the probability that when he selects two shoes at random they can be worn as a matching pair.

7 A bag contains five 4cm nails, six 7cm nails and nine 10cm nails. Find the probability that two randomly selected nails:

   a have a total length of 14cm

   b are both 7cm long, given that they have a total length of 14cm.

8 Yvonne and Novac play two games of tennis every Saturday. Yvonne has a 65% chance of winning the first game and, if she wins it, her chances of winning the second game increase to 70%. However, if she loses the first game, then her chances of winning the second game decrease to 55%. Find the probability that Yvonne:

   a loses the second game

   b wins the first game, given that she loses the second game.

9 a Given $P(X \cap Y) = 0.13$ and $P(X) = 0.65$, find $P(Y \mid X)$.

   b Given $P(M \cap N) = 0.27$ and $P(N \mid M) = 0.81$, find $P(M)$.

   c Given $P(V \cap W) = 0.35$ and $P(W') = 0.60$, find $P(V \mid W)$.

10 When a customer at a furniture store makes a purchase, there is a 15% chance that they purchase a bed. Given that 4.2% of all customers at the store purchase a bed, find the probability that a customer does not make a purchase at the store.

11 A number between 10 and 100 inclusive is selected at random. Find the probability that the number is a multiple of 5, given that none of its digits is a 5.

12 Three of Mr Jumbillo's seven children, who include one set of twins, are selected at random.

   a Calculate the probability that exactly one of the twins is selected.

   b Given that exactly three of the children are girls, find the probability that the selection of three children contains more girls than boys.

13 Anya calls Zara once each evening before she goes to bed. She calls Zara's mobile phone with probability 0.8 or her landline. The probability that Zara answers her mobile phone is 0.74, and the probability that she answers her landline is y. This information is displayed in the tree diagram shown.

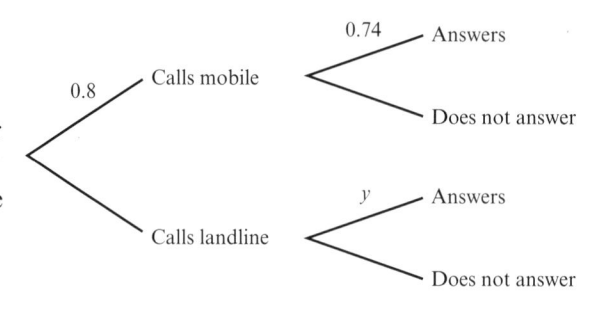

   a Given that Zara answers 68% of Anya's calls, find the value of y.

   b Given that Anya's call is not answered, find the probability that it is made to Zara's landline.

14 Every Friday, Arif offers to take his sons to the beach or to the park. The sons refuse an offer to the beach with probability 0.65 and accept an offer to the park with probability 0.85. The probability that Arif offers to take them to the beach is x. This information is shown in the tree diagram.

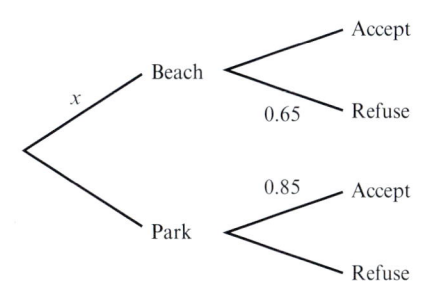

    a Find the value of x, given that 33% of Arif's offers are refused.

    b Given that Arif's offer is accepted, find the probability that he offers to take his sons to the park.

PS 15 Two children are selected from a group in which there are 10 more boys than girls. Given that there are 756 equiprobable ordered selections that can occur, find the probability that two boys or two girls are selected.

PS 16 There is a 43% chance that Riya meets her friend Jasmine when she travels to work. Given that Riya walks to work and does not meet Jasmine 30% of the time, and that she travels to work by a different method and meets Jasmine 25% of the time, find the probability that Riya walks to work on any particular day.

PS 17 Aaliyah buys a randomly selected magazine that contains a crossword puzzle on five randomly chosen days of each week. On 84% of the occasions that she buys a magazine, she attempts its crossword, which she manages to complete 60% of the time. Find the probability that, on any particular day, Aaliyah does not complete the crossword in a magazine.

> **FAST FORWARD**
>
> We will study further techniques for calculating probabilities using permutations and combinations in Chapter 5, Section 5.4.

Checklist of learning and understanding

- Probabilities are assigned on a scale from 0 (impossible) to 1 (certain).
- When one object is randomly selected from n objects, P(selecting any particular object) $= \frac{1}{n}$.
- $P(\text{event}) = \frac{\text{Number of favourable equally likely outcomes}}{\text{Total number of equally likely outcomes}}$
- $P(A) + P(\text{not } A) = 1$ or $P(A) + P(A') = 1$
- In n trials, event A is expected to occur $n \times P(A)$ times.
- $A \cup B$ means 'A or B' and $A \cap B$ means 'A and B'.
- Mutually exclusive events have no common favourable outcomes.
  For mutually exclusive events A and B, $P(A \text{ or } B) = P(A \cup B) = P(A) + P(B)$
- Non-mutually exclusive events have at least one common favourable outcome.
  For any two events A and B, $P(A \text{ or } B) = P(A \cup B) = P(A) + P(B) - P(A \cap B)$
- Independent events can occur without being affected by the occurrence of each other.
  Events A and B are independent if and only if $P(A \text{ and } B) = P(A \cap B) = P(A) \times P(B)$, such that $P(A \mid B) = P(A \mid B')$.
- For any two events A and B, $P(A \text{ and } B) = P(A \cap B) = P(A) \times P(B \mid A)$ and $P(B \mid A) = \frac{P(A \cap B)}{P(A)}$.

END-OF-CHAPTER REVIEW EXERCISE 4

1. Four quality-control officers were asked to test 1214 randomly selected electronic components from a company's production line, and to report the proportions that they found to be defective. The proportions reported were: $\frac{2}{333}, \frac{3}{411}, \frac{0}{187}$ and $\frac{4}{283}$.

   These figures confirm what the manager thought; that about $k\%$ of the components produced are defective.

   Of the 7150 components that will be produced next month, approximately how many does the manager expect to be defective? [2]

2. Three referees are needed at an international tournament and there are 12 to choose from: three from Bosnia, four from Chad and five from Denmark. If the referees are selected at random, find the probability that at least two of them are from the same country. [3]

3. The diagram opposite gives details about a company's 115 employees. For example, it employs four unqualified, part-time females.

   Two employees are selected at random. Find the probability that:

   a one is a qualified male and the other is an unqualified female [2]

   b both are unqualified, given that neither is employed part-time. [3]

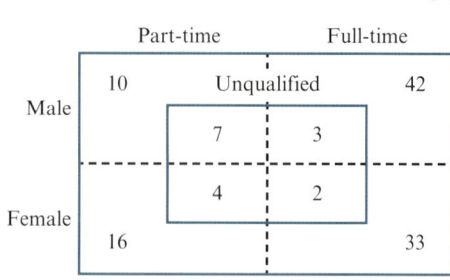

4. The numbers of books read in the past 3 months by the members of a reading club are shown in the following table.

No. books	<2	2–4	5–7	8–9	10	>10
No. members (f)	3	5	22	7	2	1

   Find the probability that three randomly selected members have all read fewer than eight books, given that they have all read more than four books. [3]

5. When they are switched on, certain small devices independently produce outputs of 1, 2 or 3 volts with respective probabilities of 0.3, 0.6 and 0.1. Find the probability that three of these devices produce an output with a sum of 5 or 6 volts. [4]

6. One hundred people are attending a conference. The following Venn diagram shows how many are male (M), have brown eyes (BE) and are right-handed (RH).

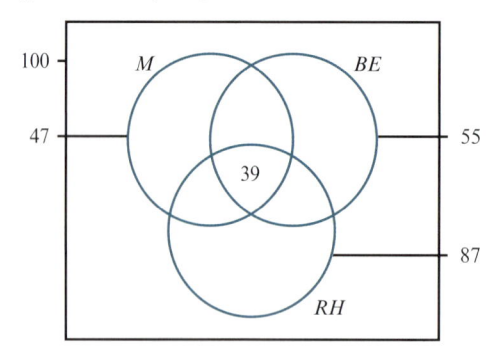

   a Given that there are 43 males with brown eyes, 42 right-handed males and 46 right-handed people with brown eyes, copy and complete the Venn diagram. [3]

   b Two attendees are selected at random. Find the probability that:

   i   they are both females who are not right-handed [2]

   ii exactly one of them is right-handed, given that neither of them have brown eyes. [3]

7 A student travels to college by either of two routes, A or B. The probability that they use route A is 0.3, and the probability that they are passed by a bus on their way to college on any particular day is 0.034. They are twice as likely to be passed by a bus when they use route B as when they use route A.

   a Use the tree diagram opposite to form and solve a pair of simultaneous equations in x and y. [3]

   b Find the probability that the student uses route B, given that they are not passed by a bus on their way to college. [3]

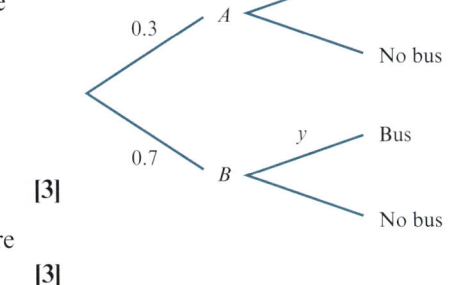

8 Two ordinary fair dice are rolled. If the first shows a number less than 3, then the score is the mean of the numbers obtained; otherwise the score is equal to half the absolute (non-negative) difference between the numbers obtained. Find the probability that the score is:

   a positive [3]

   b greater than 1, given that it is less than 2 [3]

   c less than 2, given that it is greater than 1. [3]

9 Three friends, Rick, Brenda and Ali, go to a football match but forget to say which entrance to the ground they will meet at. There are four entrances, A, B, C and D. Each friend chooses an entrance independently.

   • The probability that Rick chooses entrance A is $\frac{1}{3}$. The probabilities that he chooses entrances B, C or D are all equal.
   • Brenda is equally likely to choose any of the four entrances.
   • The probability that Ali chooses entrance C is $\frac{2}{7}$ and the probability that he chooses entrance D is $\frac{3}{5}$. The probabilities that he chooses the other two entrances are equal.

   i Find the probability that at least 2 friends will choose entrance B. [4]

   ii Find the probability that the three friends will all choose the same entrance. [4]

   Cambridge International AS & A Level Mathematics 9709 Paper 61 Q5 November 2010

10 Maria chooses toast for her breakfast with probability 0.85. If she does not choose toast then she has a bread roll. If she chooses toast then the probability that she will have jam on it is 0.8. If she has a bread roll then the probability that she will have jam on it is 0.4.

   i Draw a fully labelled tree diagram to show this information. [2]

   ii Given that Maria did **not** have jam for breakfast, find the probability that she had toast. [4]

   Cambridge International AS & A Level Mathematics 9709 Paper 62 Q3 November 2009

11 Ronnie obtained data about the gross domestic product (GDP) and the birth rate for 170 countries. He classified each GDP and each birth rate as either 'low', 'medium' or 'high'. The table shows the number of countries in each category.

		Birth rate		
		Low	Medium	High
GDP	Low	3	5	45
	Medium	20	42	12
	High	35	8	0

One of these countries is chosen at random.

  i Find the probability that the country chosen has a medium GDP. [1]

  ii Find the probability that the country chosen has a low birth rate, given that it does not have a medium GDP. [2]

  iii State with a reason whether or not the events 'the country chosen has a high GDP' and 'the country chosen has a high birth rate' are exclusive. [2]

One country is chosen at random from those countries which have a medium GDP and then a different country is chosen at random from those which have a medium birth rate.

  iv Find the probability that both countries chosen have a medium GDP and a medium birth rate. [3]

Cambridge International AS & A Level Mathematics 9709 Paper 63 Q3 November 2012

12 Three boxes, A, B and C, each contain orange balls and blue balls, as shown.

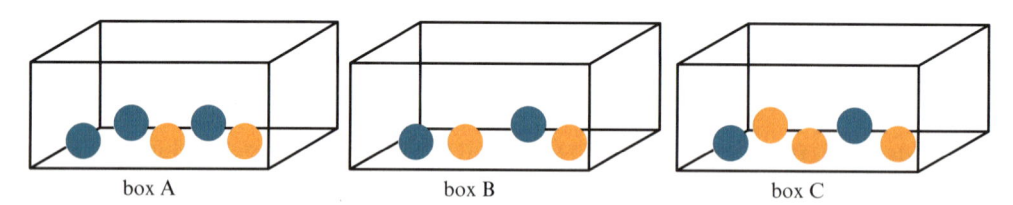

box A   box B   box C

  a A girl selects a ball at random from a randomly selected box. Given that she selects a blue ball, find the probability that it is from box C. [3]

  b A boy randomly selects one ball from each box. Given that he selects exactly one blue ball, find the probability that it is from box A. [4]

13 A and B are independent events. If $P(A) = 0.45$ and $P(B) = 0.64$, find $P[(A \cup B)']$. [2]

14 Two ordinary fair dice are rolled.

Event A is 'the sum of the numbers rolled is 2, 3 or 4'.

Event B is 'the absolute difference between the numbers rolled is 2, 3 or 4'.

  a When the dice are rolled, they show a 1 and a 3. Explain why this result shows that events A and B are not mutually exclusive. [1]

  b Explain how you know that $P(A \cap B) = \dfrac{1}{18}$. [2]

  c Determine whether events A and B are independent. [3]

15 A and B are independent events, where $P(A \cap B') = 0.14$, $P(A' \cap B) = 0.39$ and

  $P(A \cap B) < 0.25$. Use a Venn diagram, or otherwise, to find $P[(A \cup B)']$. [4]

16 In a survey, adults were asked to answer yes or no to the question 'Do you regularly watch the evening TV news?' Some of the results from the survey are detailed in the Venn diagram opposite.

One adult is selected at random and it is found that the events 'a female is selected' and 'a person who regularly watches the evening TV news is selected' are independent. Find the number of adults questioned in the survey. [4]

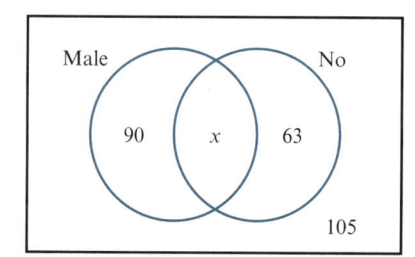

17 Bookings made at a hotel include a room plus any meal combination of breakfast (B), lunch (L) and supper (S). The Venn diagram opposite shows the number of each type of booking made by 71 guests on Friday.
   a A guest who has not booked all three meals is selected at random. Find the probability that this guest:
      i   has booked breakfast or supper [2]
      ii has not booked supper, given that they have booked lunch. [2]
   b Find the probability that two randomly selected guests have both booked lunch, given that they have both booked at least two meals. [3]

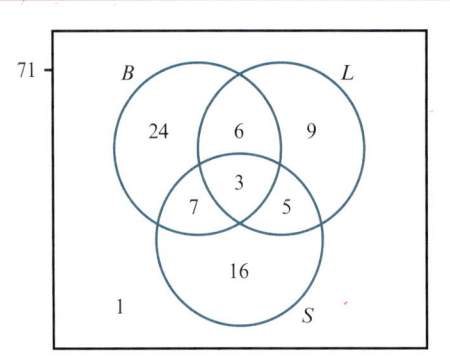

18 Three strangers meet on a train. Assuming that a person is equally likely to be born in any of the 12 months of the year, find the probability that at least two of these three people were born in the same month of the year. [4]

19 A box contains three black and four white chess pieces. Find the probability that a random selection of five chess pieces, taken one at a time without replacement, contains exactly two black pieces which are selected one immediately after the other. [4]

20 The following table shows the numbers of IGCSE (I) and A Level (A) examinations passed by a group of university students.

		IGCSEs (I)				
		5	6	7	8	9	10
A Levels (A)	2	13	9	5	4	2	0
	3	7	8	9	6	1	0
	4	0	0	1	0	1	0
	5	0	0	0	0	0	1

   a For a student selected at random, find:
      i $P(I + A = 11 \mid A < 4)$ [2]
      ii $P(I - A > 5 \mid I + A > 10)$ [3]
   b Six students who all have at least three A Level passes are selected at random. Find the greatest possible range of the total number of IGCSE passes that they could have. [2]

Chapter 5
Permutations and combinations

In this chapter you will learn how to:

- solve simple problems involving selections
- solve problems about arrangements of objects in a line, including those involving repetition and restriction
- evaluate probabilities by calculation using permutations or combinations.

Chapter 5: Permutations and combinations

PREREQUISITE KNOWLEDGE

Where it comes from	What you should be able to do	Check your skills
Chapter 4, Section 4.4.	Recognise and calculate conditional probabilities.	An experiment has 10 equally likely outcomes: three are favourable to event A, five are favourable to event B and four are favourable to neither A nor B. Find $P(A \mid B)$ and $P(B \mid A)$.

Simple situations with millions of possibilities

This topic is concerned with **selections** and **arrangements** of objects. **Permutations** and **combinations** appear in many complex modern applications: transport logistics; relationships between proteins in genetic engineering; sorting algorithms in computer science; and protecting computer passwords and e-commerce transactions in cryptography.

A selection of objects is called a combination if the order of selection does not matter; however, if the order of selection does matter, then the selection is called a permutation.

To make the difference between a permutation and a combination clear, we can describe them as follows.

- A combination is a way of selecting objects.
  There are three combinations of two letters from A, B and C. These are A and B, A and C, and B and C.
- A permutation is a way of selecting objects and arranging them in a particular order.
  There are six permutations of two letters from A, B and C. These are
  AB, BA, AC, CA, BC and CB.

EXPLORE 5.1

Each letter A to Z is encrypted (or transformed) to a fixed distinct letter using its position in the alphabet ($A = 1, B = 2, C = 3,\ldots$). By doing this, the password SATURN is encrypted as ECHKBP.

This information gives us six clues to work out the method of encryption (e.g. S → E means 19 → 5, A → C means 1 → 3, and so on).

An Enigma Encryption Machine, circa 1940.

Investigate the method of encryption and then find the password that is encrypted as UJSNOL.

There are over 27 million possible passwords, so the probability that a random guess is correct is approximately zero.

5.1 The factorial function

In this chapter, we will frequently need to write and evaluate expressions such as $4 \times 3 \times 2 \times 1$.

A shorthand method of doing this is to use the **factorial** function: $4 \times 3 \times 2 \times 1$ is called 'four factorial' and is written 4! On most calculators, the factorial function appears as $n!$ or $x!$.

As examples:

$7! = 7 \times 6 \times 5 \times 4 \times 3 \times 2 \times 1 = 5040$

$\dfrac{6!}{4!} = \dfrac{6 \times 5 \times 4 \times 3 \times 2 \times 1}{4 \times 3 \times 2 \times 1} = 6 \times 5 = 30$

$5! - 4! = 4!(5 - 1) = 4! \times 4 = 96$

$\dfrac{3!}{0!} = \dfrac{3 \times 2 \times 1}{1} = 6$

KEY POINT 5.1

$n! = n(n-1)(n-2)\ldots \times 3 \times 2 \times 1$, for any integer $n > 0$. $0! = 1$

The following figure shows values of $n!$ in sequence, where the next term is obtained by division. From this it is clear that 0! must be equal to 1.

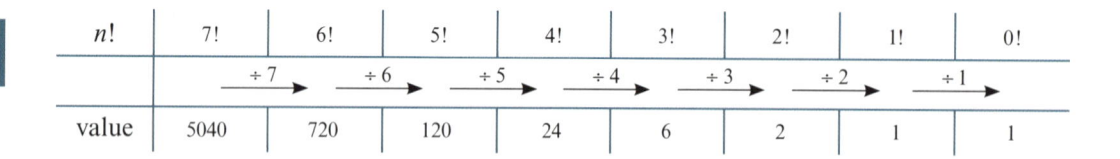

$n!$	7!	6!	5!	4!	3!	2!	1!	0!
	÷7	÷6	÷5	÷4	÷3	÷2	÷1	
value	5040	720	120	24	6	2	1	1

EXERCISE 5A

1 Without using a calculator, find the value of:

   a $\dfrac{5!}{3!}$
   b $\dfrac{4!}{2!} - 3!$
   c $7 \times 4! + 21 \times 3!$
   d $\dfrac{10!}{8!} + \dfrac{9!}{7!}$
   e $\dfrac{20!}{18!} - \dfrac{13!}{11!}$

2 Use your calculator to find the smallest value of n for which:

   a $n! > 1\,000\,000$
   b $5! \times 6! < n!$
   c $(n!)! > 10^{20}$

3 Use your calculator to find the largest value of n for which:

   a $\dfrac{n!}{500\,000} < 80$
   b $1.5 \times 10^{12} - n! > 0$
   c $\dfrac{n!}{(n-2)!} < 500$

4 Express, in as many different ways as possible, the numbers 144, 252 and $1\tfrac{1}{2}$ in the form $\dfrac{a! \times b!}{c!}$, where none of a, b or c is equal to 0 or to 1.

5 Express the area of a 53 cm by 52 cm rectangle using factorials.

6 Two cubical boxes measure 25 cm by 24 cm by 23 cm, and 8 cm by 7 cm by 6 cm. Express the difference between their volumes using factorials.

7 Eight children each have seven boxes of six eggs and each egg is worth $0.09. Write the total value of all these eggs in dollars, using factorials.

Chapter 5: Permutations and combinations

DID YOU KNOW?

There is a famous legend about the Grand Vizier in Persia who invented chess.

The King was so delighted with the new game that he invited the Vizier to name his own reward. The Vizier replied that, being a modest man, he wanted only one grain of wheat on the first square of a chessboard, two grains on the second, four on the third, and so on, with twice as many grains on each square as on the previous square. The innumerate King agreed, not realising that the total number of grains on all 64 squares would be $2^{64} - 1$, or 1.84×10^{19}, which is equivalent to the world's present wheat production for the next 150 years.

Although the number $2^{64} - 1$ is extremely large, it is only about one-third of 21 factorial.

As a challenge, try showing that $2^0 + 2^1 + 2^2 + ... + 2^{63} = 2^{64} - 1$ without using a formula.

5.2 Permutations

We can make a permutation by taking a number of objects and arranging them in a line. For example, the two possible permutations of the digits 5 and 9 are the numbers 59 and 95.

Although there are several methods that we can use to find the number of possible permutations of objects, all methods involve use of the factorial function.

Permutations of n distinct objects

The number of permutations of n distinct objects is denoted by ^nP_n, and there are $n!$ permutations that can be made. For example, there are $^2P_2 = 2! = 2$ permutations of the two digits 5 and 9, as we have just seen.

KEY POINT 5.2

The number of permutations of n distinct objects is $^nP_n = n! = n(n-1)(n-2)... \times 3 \times 2 \times 1$, for any integer $n > 0$.

Consider all the three-digit numbers that can be made by arranging the digits 5, 6 and 7.

In this simple case, we can make a list to show there are six possible three-digit numbers.

These are 567, 576, 657, 675, 756 and 765.

The following tree diagram gives another method of showing the six possible arrangements of the three digits.

left	middle	right	number
5	6	7	567
5	7	6	576
6	5	7	657
6	7	5	675
7	5	6	756
7	6	5	765

Unfortunately, writing out lists and constructing tree diagrams to find numbers of possible arrangements of objects are suitable methods only for small numbers of objects. Imagine listing all the possible arrangements of seven different letters; there would be over 5000 on the list and a tree diagram would have over 5000 branches at its right-hand side!

Clearly, a more practical method for finding numbers of arrangements is needed. This is the primary use of the factorial function.

We can show that six three-digit numbers can be made from 5, 6 and 7 by considering how many choices we have for the digit that we place in each position in the arrangement. If we first place a digit at the left side, we have three choices. Next, we place a digit in the middle (two choices). Finally, we place the remaining digit at the right side, as shown in the following diagram.

$$\underline{3}_{\text{choices}} \times \underline{2}_{\text{choices}} \times \underline{1}_{\text{choices}}$$

The numbers above the lines in the diagram are not the digits that are being arranged – they are the numbers of choices that we have for placing the three digits.

The three digits can be arranged in $\underline{3} \times \underline{2} \times \underline{1} = 3! = {}^3P_3 = 6$ ways.

The seven letters mentioned previously can be arranged in
$\underline{7} \times \underline{6} \times \underline{5} \times \underline{4} \times \underline{3} \times \underline{2} \times \underline{1} = 7! = 5040$ ways.

WORKED EXAMPLE 5.1

In how many different ways can five boys be arranged in a row?

Answer

$\underline{5} \times \underline{4} \times \underline{3} \times \underline{2} \times \underline{1} = 5! = {}^5P_5 = 120$ ways.

We multiply together the number of choices for each of the five positions, working from left to right.

TIP

We could just as easily work from right to left, giving $\underline{1} \times \underline{2} \times \underline{3} \times \underline{4} \times \underline{5} = 5!$ ways.

WORKED EXAMPLE 5.2

In how many ways can nine elephants and four mice be arranged in a line?

Answer

$13! = {}^{13}P_{13} = 6\,227\,020\,800$ ways.

The nine elephants and four mice are distinct, so we are arranging 13 different animals.

TIP

Large number answers can be given more accurately than to 3 significant figures.

Chapter 5: Permutations and combinations

EXERCISE 5B

1. In how many ways can the six letters A, B, C, D, E and F be arranged in a row?

2. From a standard deck of 52 playing cards, find how many ways there are of arranging in a row:
   - **a** all 52 cards
   - **b** the four kings
   - **c** the 13 diamonds.

3. In how many different ways can the following stand in a line?
   - **a** two women
   - **b** six men
   - **c** eight adults.

4. In how many different ways can the following sit in a row on a bench?
   - **a** four girls
   - **b** three boys
   - **c** four girls and three boys.

5. Seven cars and x vans can be parked in a line in 39 916 800 ways. Find the number of ways in which five cars and $x + 2$ vans can be parked in a line.

6. A woman has 10 children. She arranges 11 chairs in a row and sits on the chair in the middle. If her youngest child sits on the adjacent chair to her left, in how many ways can the remaining children be seated?

7. **PS** A group of n boys can be arranged in a line in a certain number of ways. By adding two more boys to the group, the number of possible arrangements increases by a factor of 420. Find the value of n.

Permutations of n objects with repetitions

When n objects include repetitions (i.e. when they are not all distinct), there will be fewer than ^nP_n permutations, so an adjustment to the use of the factorial function is needed.

Consider making five-letter arrangements with A, A, B, C and D.

To simplify the problem, we can distinguish the repeated A by writing the letters as A, A, B, C and D.

A**A**BCD is the same arrangement as **A**ABCD.

D**A**CAB is the same arrangement as DAC**A**B, and so on.

Each time we swap A and **A**, we obtain the same arrangement.

If the five letters were distinct, there would be $^5P_5 = 5! = 120$ arrangements, but the number is reduced to half $\left(\dfrac{1}{2!}\right)$ of the total arrangements because the two repeated letters can be placed in 2! ways in any particular arrangement without changing that arrangement.

Letters to be arranged = 5; number of the same letter = 2.

There are $\dfrac{^5P_5}{^2P_2} = \dfrac{5!}{2!} = 60$ five-letter arrangements that can be made.

> **KEY POINT 5.3**
>
> The number of permutations of n objects, of which p are of one type, q are of another type, r are of another type, and so on, is $\dfrac{^nP_n}{p! \times q! \times r! \times \ldots} = \dfrac{n!}{p! \times q! \times r! \times \ldots}$, where $p + q + r + \ldots = n$.

WORKED EXAMPLE 5.3

The capital of Burkina Faso is OUAGADOUGOU. Find the number of distinct arrangements of all the letters in this word.

Answer

$\dfrac{11!}{3! \times 3! \times 2! \times 2!} = 277\,200$ 11 letters are to be arranged, with repeats of three Os, three Us, two As and two Gs. In the formula of Key point 5.3, excluding 1! for the D in the denominator does not change our answer.

EXERCISE 5C

1 Find the number of distinct arrangements of all the letters in these words:
   - a TABLE
   - b TABLET
   - c COMMITTEE
   - d MISSISSIPPI
   - e HULLABALLOO.

2 Find how many six-digit numbers can be made from these sets of digits:
   - a 1, 1, 1, 1, 1 and 3
   - b 2, 2, 2, 7, 7 and 7
   - c 5, 6, 6, 6, 7 and 7
   - d 8, 8, 9, 9, 9 and 9.

3 A girl has 20 plastic squares. There are five identical red squares, seven identical blue squares and eight identical green squares. By placing them in a row, joined edge-to-edge, find how many different arrangements she can make using:
   - a one square of each colour
   - b the five red squares only
   - c all of the blue and green squares
   - d all of the 20 squares.

4 Two students are asked to find how many ways there are to plant two trees and three bushes in a row. The first student gives $5! = 120$, and the second gives $\dfrac{5!}{2! \times 3!} = 10$. Decide who you agree with and explain the error made by the other student.

5 Ten coins are placed in a row on a table, each showing a head or a tail.
   - a How many different arrangements of heads and/or tails are possible?
   - b Of the arrangements in part **a**, find how many have:
      - i five heads and five tails showing
      - ii more heads than tails showing.

6 There are 420 possible arrangements of all the letters in a particular seven-letter word. Give a description of the letters in this word.

7 Find the number of distinct five-letter arrangements that can be made from:
   - a two As and three Bs
   - b two identical vowels and three Bs
   - c two identical vowels and any three identical consonants.

Chapter 5: Permutations and combinations

EXPLORE 5.2

Consider the number of distinct arrangements of the 16 letters in the word COUNTERCLOCKWISE – there are close to 8.72×10^{11}. We rarely meet such numbers in our daily lives, so we are likely to see this as just a *very large number* whose true size we cannot really comprehend until it is put into some human context. For example, if everyone on Earth over the age of 14 (i.e. about 5.46×10^9 people) contributed one new arrangement of the word every day starting on 1st January, we would complete the list of arrangements around 9th June.

The calculation for this is $\dfrac{8.72 \times 10^{11}}{5.46 \times 10^9} \approx 160$ days.

Devise a way of expressing the number of distinct arrangements of the letters in the word PNEUMONOULTRAMICROSCOPICSILICOVOLCANOCONIOSIS (which is the full name for the disease known as silicosis, and is the longest word in any major English language dictionary) in a way that is meaningful to human understanding.

There are, for example, 3.15×10^7 seconds in a year, about 7.48×10^9 people on Earth, and the masses of the Earth and Sun are 5.97×10^{24} and 1.99×10^{30} kg, respectively.

WEB LINK

For other options, perform a web search for *large numbers*.

Permutations of *n* distinct objects with restrictions

The number of possible arrangements of objects is reduced when restrictions are put in place. As a general rule, the number of choices for the restricted positions should be investigated first, and then the unrestricted positions can be attended to.

WORKED EXAMPLE 5.4

Find the number of ways of arranging six men in a line so that:

a the oldest man is at the far-left side

b the two youngest men are at the far-right side

c the shortest man is at neither end of the line.

Answer

Without restrictions, the six men can be arranged in $^6P_6 = 6! = 720$ ways. So, with restrictions, there will be fewer than 720 arrangements.

a $1 \times {}^5P_5 = 1 \times 5! = 120$ arrangements

The oldest man must be at the far-left side (one choice), and the other five men can be arranged in the remaining five positions in $^5P_5$ ways.

b $\underbrace{4 \times 3 \times 2 \times 1}_{^4P_4} \times \underbrace{2 \times 1}_{^2P_2}$

$^4P_4 \times {}^2P_2 = 4! \times 2! = 48$ arrangements

The two spaces at the right are reserved for the two youngest men, who can be placed there in $^2P_2$ ways. The other four men can be arranged in the remaining four positions in $^4P_4$ ways, as shown.

Cambridge International AS & A Level Mathematics: Probability & Statistics 1

c $\underbrace{5 \times \underbrace{4 \times 3 \times 2 \times 1}_{^4P_4} \times 4}$

There are only five men who can be placed at the far-left side, so there are only four men who can be placed at the far right. The remaining four positions can be filled by any of the other four men (one of whom is the shortest man) in $^4P_4$ ways, as shown.

$5 \times {}^4P_4 \times 4 = 5 \times 4! \times 4 = 480$ arrangements.

Alternatively, the shortest man can be placed in one of four positions, and the other five positions can be filled in $^5P_5$ ways, so $4 \times 5! = 480$.

WORKED EXAMPLE 5.5

How many odd four-digit numbers greater than 3000 can be made from the digits 1, 2, 3 and 4, each used once?

Answer

Restrictions affect the digits in the thousands column and in the units column. The digit at the far left (i.e. thousands column) can be only 3 or 4, and the digit at the far right (i.e. units column) can be only 1 or 3. The 3 can be placed in either of the restricted positions, so we can investigate separately the four-digit numbers that start with 3, and the four-digit numbers that start with 4.

$\underbrace{1 \times \underbrace{2 \times 1}_{^2P_2} \times 1}$

$1 \times {}^2P_2 \times 1 = 2$ numbers

Start with 3: We must place 1 at the far right (one choice), and the remaining two positions can be filled by the other two digits in $^2P_2$ ways, as shown.

$\underbrace{1 \times \underbrace{2 \times 1}_{^2P_2} \times 2}$

$1 \times {}^2P_2 \times 2 = 4$ numbers

Start with 4: We can place 1 or 3 at the far right (two choices), and the remaining two positions can be filled by the other two digits in $^2P_2$ ways, as shown.

$2 + 4 = 6$ odd numbers greater than 3000 can be made.

💡 **TIP**

Alternatively, we could solve this problem by investigating separately the numbers that end with 1, and the numbers that end with 3.

💡 **TIP**

These six numbers are 3241, 3421, 4123, 4213, 4231 and 4321.

WORKED EXAMPLE 5.6

Find how many ways two mangoes (M) and three watermelons (W) can be placed in a line if the five fruits are distinguishable and the mangoes:

a must not be separated   b must be separated

Answer

a $\underbrace{\overbrace{M_1 \; M_2}^{^2P_2} \; W_1 \; W_2 \; W_3}_{^4P_4}$

$\underbrace{M_1 \; M_2}_{\text{1 object}}$

$^2P_2 \times {}^4P_4 = 48$ ways

The two mangoes can be placed next to each other in $^2P_2$ ways. This pair is now considered as a single object to be arranged with the three watermelons, giving a total of four objects to arrange, as shown.

💡 **TIP**

Objects that must not be separated are treated as a single object when arranged with others.

Chapter 5: Permutations and combinations

> **b** $120 - 48 = 72$ ways
>
> With no restrictions, the five items can be arranged in $^5P_5 = 5! = 120$ ways, and we know that the mangoes are not separated in 48 of these.

EXERCISE 5D

1. Find how many five-digit numbers can be made using the digits 2, 3, 4, 5 and 6 once each if:
   a. there are no restrictions
   b. the five-digit number must be:
      i. odd
      ii. even
      iii. odd and less than 40 000.

2. Find how many ways four men and two women can stand in a line if:
   a. the two women must be at the front
   b. there must be a woman at the front and a man at the back
   c. the two women must be separated
   d. the four men must not be separated
   e. no two men may stand next to each other.

3. Find the ratio of odd-to-even six-digit numbers that can be made using the digits 1, 2, 3, 4, 5 and 7.

4. Find how many ways 10 books can be arranged in a row on a shelf if:
   a. the two oldest books must be in the middle two positions
   b. the three newest books must not be separated.

5. Five cows and one set of twin calves can be housed separately in a row of seven stalls in $^7P_7 = 5040$ ways. Find in how many of these arrangements:
   a. the two calves are not in adjacent stalls
   b. the two calves and their mother, who is one of the 5 cows, are in adjacent stalls
   c. each calf is in a stall adjacent to its mother.

6. Find how many of the six-digit numbers that can be made from 1, 2, 2, 3, 3 and 3:
   a. begin with a 2
   b. are not divisible by 2.

7. Find the number of distinct arrangements that can be made from all the letters in the word THEATRE when the arrangement:
   a. begins with two Ts and ends with two Es
   b. has H as its middle letter
   c. ends with the three vowels E, A and E.

8. The following diagram shows a row of post boxes with the owners' names beneath.

   Five parcels, one for the owner of each box, have arrived at the post office. If one parcel is randomly placed in each box, find the number of ways in which:

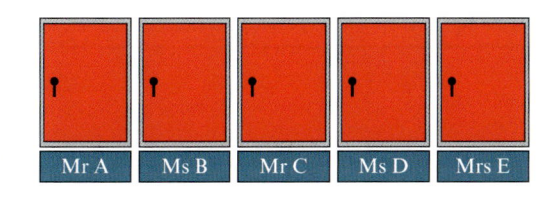

a the five parcels can all be placed in the correct boxes

b exactly one parcel can be placed in the wrong box

c the correct parcels can be placed in Mr A's and one other person's box only

d exactly two parcels can be placed in the correct boxes.

P 9 There are x boys and y girls to be arranged in a line. Find the relationship between x and y if it is not possible to separate all the boys.

Permutations of r objects from n objects

So far, we have dealt only with permutations in which all of the objects are selected and arranged. We can now take this a step further and look at permutations in which only some of the objects are selected and arranged. When we select and arrange r objects in a particular order from n distinct objects, we call this a permutation of r from n.

Suppose, for example, we wish to select and arrange three letters from the five letters A, B, C, D and E.

We have five choices for the first letter, four for the second, and three choices for the third. This gives us a total of $\underline{5} \times \underline{4} \times \underline{3} = 60$ permutations, which is effectively 5! but 2! are missing. There are $\dfrac{5!}{(5-3)!} = \dfrac{5!}{2!} = 60$ permutations altogether.

> **KEY POINT 5.4**
>
> There are $^nP_r = \dfrac{n!}{(n-r)!}$ permutations of r objects from n distinct objects.

WORKED EXAMPLE 5.7

How many three-digit numbers can be made from the seven digits 3, 4, 5, 6, 7, 8 and 9, if each is used at most once?

Answer

$^7P_3 = \dfrac{7!}{(7-3)!}$

$= \dfrac{7!}{4!}$

$= 7 \times 6 \times 5$

$= 210$ three-digit numbers

We select and arrange just three of the seven distinct digits (and ignore four of them).

> **TIP**
>
> The choices we have for the first, second and third digits are $\underline{7} \times \underline{6} \times \underline{5} = 210$

WORKED EXAMPLE 5.8

In how many ways can five playing cards from a standard deck of 52 cards be arranged in a row?

Answer

$^{52}P_5 = \dfrac{52!}{47!}$

$= 311\,875\,200$ ways

We select and arrange five of the 52 playing cards (and ignore 47 of them).

> **TIP**
>
> The choices we have are $\underline{52} \times \underline{51} \times \underline{50} \times \underline{49} \times \underline{48}$
> $= 311\,875\,200$.

Chapter 5: Permutations and combinations

WORKED EXAMPLE 5.9

In how many ways can 4 out of 18 girls sit on a four-seat sofa when the oldest girl must be given one of the seats?

Answer

$4 \times {}^{17}P_3 = 16320$ ways

Four ways for the oldest girl to occupy a seat, and ${}^{17}P_3$ ways to select and arrange three of the remaining 17 girls to sit with her.

WORKED EXAMPLE 5.10

In how many ways can four boys and three girls stand in a row when no two girls are allowed to stand next to each other?

Answer

${}^4P_4$ ways to arrange 4 boys in a row

↑ B_1 ↑ B_2 ↑ B_3 ↑ B_4 ↑

Arrange the girls in 3 of these 5 spaces ${}^5P_3$

${}^4P_4 \times {}^5P_3 = 1440$ ways

${}^4P_4$ ways to arrange the four boys in a row.

${}^5P_3$ ways to select three of the five spaces between or to the side of the boys and arrange the three girls in them, as shown.

TIP

Objects that *must* be separated are individually placed between or beyond the objects that *can* be separated.

EXERCISE 5E

1. Find how many permutations there are of:

    a five from seven distinct objects
    b four from nine distinct objects.

2. From 12 books, how many ways are there to select and arrange exactly half of them in a row on a shelf?

3. In how many ways can gold, silver and bronze medals be awarded for first, second and third places in a race between 20 athletes? You may assume that no two athletes tie in these positions.

4. a Find the number of ways in which Alvaro can paint his back door and his front door in a different colour if he has 14 colours of paint to choose from.

    b In how many ways could Alvaro do this if he also considered painting them the same colour?

5. Find how many of the arrangements of four letters from A, B, C, D, E and F:

    a begin with the letter A
    b contain the letter A.

6. From a group of 10 boys and seven girls, two are to be chosen to act as the hero and the villain in the school play. Find in how many ways this can be done if these two roles are to be played by:

    a any of the children
    b two girls or two boys
    c a boy and a girl.

Cambridge International AS & A Level Mathematics: Probability & Statistics 1

7 From a set of 10 rings, a jeweller wishes to display seven of them in their shop window. The formation of the display is shown in the diagram opposite.

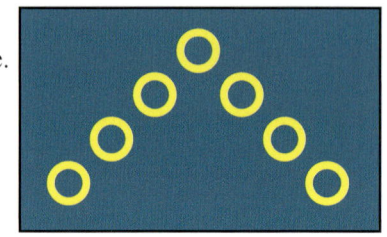

Find the number of possible displays if, from the set of 10:

a the ring with the largest diamond must go at the top of the display

b the most expensive ring must go at the top with the two least expensive rings adjacent to it.

8 Using each digit not more than once, how many even four-digit numbers can be made from the digits 1, 2, 3, 4, 5, 6 and 7?

9 Find how many three-digit numbers can be made from the digits 0, 1, 2, 3 and 4, used at most once each, if the three-digit number:

a must be a multiple of 10   b cannot begin with zero.

10 Give an example of a practical situation where the calculation $^nP_r = 120$ might arise.

P 11 a Under what condition is $^nP_r > {}^nP_{n-r}$?

b Given that $^nP_r \times {}^nP_{n-r} = k \times {}^nP_n$, find an expression for k in terms of n and r.

12 Five playing cards are randomly selected from a standard deck of 52 cards. These five cards are shuffled, and then the top three cards are placed in a row on a table. How many different arrangements of three of the 52 cards are possible?

PS 13 Seven chairs, A to G, are arranged as shown.

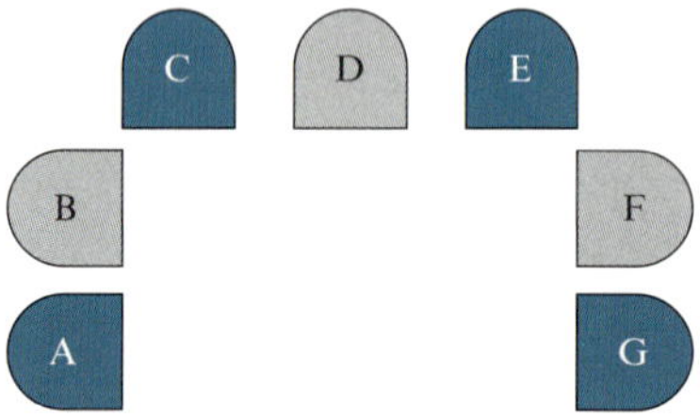

In how many ways can the chairs be occupied by 7 of a group of 12 people if three particular people are asked to sit on chairs B, D and F, in any order?

PS 14 A minibus has 11 passenger seats. There are six seats in a row on the sunny side and five seats in a row on the shady side, as shown in the following diagram.

Find how many ways eight passengers can be arranged in these seats if:

a there are no restrictions

b one particular passenger refuses to sit on the sunny side

c two particular passengers refuse to sit in seats that are either next to each other or one directly in front of the other.

DID YOU KNOW?

The rule to determine the number of permutations of n objects was known in Indian culture at least as early as 1150 and is explained in the *Līlāvatī* by Indian mathematician Bhaskara II.

In his books *Campanalogia* and *Tintinnalogia*, Englishman Fabian Stedman in 1677 described factorials when explaining the number of permutations of the ringing of church bells.

A complete peal of changes of n bells is made when they are rung in $n!$ sequences without repetition.

The speed at which church bells ring cannot be changed very much by the ringers and this may be why there are at most five bells in most churches. For example, 10 bells can be rung in 3 628 800 different sequences and it would take the ringers over 3 months to ring a complete peal of changes of 10 bells!

5.3 Combinations

A combination is simply a selection, where the order of selection is not important. Choosing strawberries and ice cream from a menu is the same combination as choosing ice cream and strawberries.

When we select r objects in no particular order from n objects, we call this is a combination.

A combination of r objects which are then arranged in order is equivalent to a permutation. We write ^nC_r to mean the number of combinations of r objects from n. Since there are $^rP_r = r!$ ways of arranging the r objects, we have:

$$^nC_r \times {^rP_r} = {^nP_r}$$

$$^nC_r \times r! = \frac{n!}{(n-r)!}$$

$$^nC_r = \frac{n!}{r!(n-r)!}$$

Suppose we wish to select three children from a group of five. We can view this task as 'choosing three and ignoring two' or as 'choosing to ignore two and remaining with three'. Regardless of how we view it, choosing three from five and choosing two from five can be done in an equal number of ways, and so $^5C_3 = {^5C_2}$.

The following three points should be noted.

$$^nC_r = {^nC_{n-r}}$$

$$^nC_r \leqslant {^nP_r}$$

$$^nC_r = \frac{n!}{r!(n-r)!} = \frac{\text{No. we select from!}}{\text{No. selected!} \times \text{No. not selected!}}$$

> **KEY POINT 5.5**
>
> There are
> $$^nC_r = \frac{n!}{r!(n-r)!}$$
> combinations of r objects from n distinct objects.

> **FAST FORWARD**
>
> $$\binom{n}{r} = \frac{n!}{r!(n-r)!}$$
>
> We will use this more modern notation in Chapter 7. However, most calculators use the ^nC_r notation, so we will use this in the current chapter.

WORKED EXAMPLE 5.11

In how many ways can three fish be selected from a bowl containing seven fish and two potatoes?

Answer

$^7C_3 = \dfrac{7!}{3! \times 4!} = 35$ ways   •••••• The two potatoes are irrelevant. We select three fish from seven fish.

WORKED EXAMPLE 5.12

In how many ways can five books and three magazines be selected from eight books and six magazines?

Answer

$^8C_5 = 56$ ways
$^6C_3 = 20$ ways   •••••• We select five from eight books *and* we select three from six magazines.

$^8C_5 \times {^6C_3} = 56 \times 20$
$\qquad\qquad\quad = 1120$ ways

> **TIP**
>
> The books and the magazines are selected independently, so we multiply the numbers of combinations.

Cambridge International AS & A Level Mathematics: Probability & Statistics 1

WORKED EXAMPLE 5.13

A team of five is to be chosen from six women and five men. Find the number of possible teams in which there will be more women than men.

Answer

From 6 women	From 5 men	No. teams
3	2	$^6C_3 \times {}^5C_2 = 200$
or 4	1	$^6C_4 \times {}^5C_1 = 75$
or 5	0	$^6C_5 \times {}^5C_0 = 6$

The table shows the possible make-up of the team when it has more women than men in it; and also the number of ways in which those teams can be chosen.

$200 + 75 + 6 = 281$ teams with more women than men.

FAST FORWARD

You will learn about probability distributions for the number of objects that can be selected in Chapter 6, such as the number of women selected for this team.

WORKED EXAMPLE 5.14

How many distinct three-digit numbers can be made from five cards, each with one of the digits 5, 5, 7, 8 and 9 written on it?

Answer

The 5 is a repeated digit, so we must investigate three situations separately.

No 5s selected: $^3P_3 = 6$ three-digit numbers. ...... The digits 7, 8 and 9 are selected and arranged.

One 5 selected: $^3C_2 \times 3! = 18$ three-digit numbers. ...... Two digits from 7, 8 and 9 are selected and arranged with a 5.

Two 5s selected: $^3C_1 \times \dfrac{3!}{2!} = 9$ three-digit numbers. ...... One digit from 7, 8 and 9 is selected and arranged with two 5s.

$6 + 18 + 9 = 33$ three-digit numbers can be made.

TIP

The selections in these three situations are mutually exclusive, so we add together the numbers of three-digit numbers.

EXERCISE 5F

1 Find the number of ways in which five apples can be selected from:
   a eight apples
   b nine apples and 12 oranges.

2 From seven men and eight women, find how many ways there are to select:
   a four men and five women
   b three men and six women
   c at least 13 people.

3 a How many different hands of five cards can be dealt from a standard deck of 52 playing cards?
   b How many of the hands in part **a** consist of three of the 26 red cards and two of the 26 black cards?

4 a From the 26 letters of the English alphabet, find how many ways there are to choose:
      i six different letters
      ii 20 different letters.
   b Use your results from part **a** to find the condition under which $^xC_y = {}^xC_z$, where x is a positive integer.

Chapter 5: Permutations and combinations

5   In a classroom there are four lights, each operated by a switch that has an *on* and an *off* position. How many possible lighting arrangements are there in the classroom?

6   From six boys and seven girls, find how many ways there are to select a group of three children that consists of more girls than boys.

7   A bag contains six red fuses, five blue fuses and four yellow fuses. Find how many ways there are to select:

   a   three fuses of different colours   b   three fuses of the same colour

   c   10 fuses in exactly two colours   d   nine fuses in exactly two colours.

8   The diagram opposite shows the activities offered to children at a school camp.

   If children must choose three activities to fill their day, how many sets of three activities are there to choose from?

   > **Today's Activities**
   > Morning: acting, painting or singing
   > Afternoon: swimming, tennis, golf or cricket
   > Evening: night-hike, star-gazing or drumming
   > *Afternoon swimming can be done at the pool or at the lake*

9   Two taxis are hired to take a group of eight friends to the airport. One taxi can carry five passengers and the other can carry three passengers.

   What information is given in this situation by the fact that $^8C_5 = {}^8C_3 = 56$?

10   Ten cars are to be parked in a car park that has 20 parking spaces set out in two rows of 10. Find how many different patterns of unoccupied parking spaces are possible if:

   a   the cars can be parked in any of the 20 spaces

   b   the cars are parked in the same row

   c   the same number of cars are parked in each row

   d   two more cars are parked in one row than in the other.

11   A boy has eight pairs of trousers, seven shirts and six jackets. In how many ways can he dress in trousers, shirt and jacket if he refuses to wear a particular pair of red trousers with a particular red shirt?

12   A girl has 11 objects to arrange on a shelf but there is room for only seven of them.

   In how many ways can she arrange seven of the objects in a row along the shelf, if her clock must be included?

13   A Mathematics teacher has 10 different posters to pin up in their classroom but there is enough space for only five of them. They have three posters on algebra, two on calculus and five on trigonometry. In how many ways can they choose the five posters to pin up if:

   a   there are no restrictions

   b   they decide not to pin up either of the calculus posters

   c   they decide to pin up at least one poster on each of the three topics algebra, calculus and trigonometry?

14   As discussed at the beginning of this chapter in Explore 5.1 about encrypting letters, it states that there are over 27 million possibilities for the password encrypted as UJSNOL. How many possibilities are there?

15   How many distinct three-digit numbers can be made from 1, 2, 2, 3, 4 and 5, using each at most once?

16 From three sets of twins and four unrelated girls, find how many selections of five people can be made if exactly:

    a two sets of twins must be included

    b one set of twins must be included.

EXPLORE 5.3

Two women and three men can sit on a five-seater bicycle in $5! = 120$ different ways. The photo shows an arrangement in which the two women are separated and the three men are also separated.

Consider, separately, the arrangements in which the women, and in which the men, are all separated from each other.

a Women separated from each other.	**b** Men separated from each other.
Women next to each other $= 2!$ Arrange three men with the women as a single object $= 4!$ There are $2! \times 4!$ arrangements in which the women are not separated. So there are $5! - (2! \times 4!) = 72$ arrangements in which the women are separated from each other.	Men next to each other $= 3!$ Arrange two women with the men as a single object $= 3!$ There are $3! \times 3!$ arrangements in which the men are not separated. So there are $5! - (3! \times 3!) = 84$ arrangements in which the men are separated from each other.

The calculations in **a** and **b** follow the same steps; however, the logic in one of them is flawed. Which of the two answers is correct? Can you explain why the other answer is not correct?

5.4 Problem solving with permutations and combinations

Permutations and combinations can be used to find probabilities for certain events.

If an event consists of a number of favourable permutations that are equiprobable, or a number of favourable combinations that are equiprobable, then

$$P(\text{event}) = \frac{\text{No. favourable permutations}}{\text{No. possible permutations}}$$

$$P(\text{event}) = \frac{\text{No. favourable combinations}}{\text{No. possible combinations}}$$

Using either of the previous given methods can greatly reduce the amount of working required to solve problems in probability. Nevertheless, we must decide carefully which of them, if any, it is appropriate to use.

REWIND

From the introduction to this chapter, we know that the order of selection matters in a permutation but does not matter in a combination.

WORKED EXAMPLE 5.15

There are 15 identical tins on a shelf. None of the tins are labelled but it is known that eight contain soup (S), four contain beans (B) and three contain peas (P).

If seven tins are randomly selected without replacement, find the probability that exactly five of them contain soup.

Answer

Favourable selections are when five tins of soup and two tins that are not soup are selected. (It is not important whether these two tins contain beans or peas.) We denote the 15 tins by $8S$ and $7S'$, where S' represents *not soup*.

$^8C_5 \times {^7C_2}$ favourable combinations ..... Selecting $5S$ from $8S$ and $2S'$ from $7S'$.

$^{15}C_7$ possible combinations ..... Selecting seven from 15 tins.

$$P(\text{select 5 tins of soup}) = \frac{^8C_5 \times {^7C_2}}{^{15}C_7}$$

$$= \frac{56 \times 21}{6435}$$

$$= \frac{392}{2145} \text{ or } 0.183.$$

> **TIP**
>
> In the numerator we have $8 + 7 = 15$ and $5 + 2 = 7$.

WORKED EXAMPLE 5.16

A girl has a bag containing 13 red cherries (R) and seven black cherries (B). She takes five cherries from the bag at random. Find the probability that she takes more red cherries than black cherries.

Answer

	From 13 red	From 7 black	Number of ways
	5	0	$^{13}C_5 \times {^7C_0} = 1287$
or	4	1	$^{13}C_4 \times {^7C_1} = 5005$
or	3	2	$^{13}C_3 \times {^7C_2} = 6006$
			Total = 12298

The table shows the possible make-up of the selected cherries when there are more red than black; and also the number of ways in which those cherries can be chosen.

$^{20}C_5 = 15504$ ways ..... Selecting five from 20 cherries.

$$P(\text{more red than black}) = \frac{12298}{15504} \text{ or } 0.793.$$

> **REWIND**
>
> From Chapter 4, Section 4.2, recall that $P(A \text{ or } B \text{ or } C)$
> $= P(A) + P(B) + P(C)$
> for mutually exclusive events.

EXPLORE 5.4

We can, of course, find the solution to Worked example 5.16 using conditional probabilities.

- There is one way to select $5R$ and $0B$.
- There are five ways to select $4R$ and $1B$.
- There are 10 ways to select $3R$ and $2B$.

Complete the calculations using conditional probabilities.

Note how much working is involved and how long the calculations take.

Compare the two approaches to solving this problem and decide for yourself which you prefer.

REWIND

We studied conditional probabilities in Chapter 4, Section 4.4.

WORKED EXAMPLE 5.17

A minibus has seats for the driver (D) and seven passengers, as shown.

When seven passengers are seated in random order, find the probability that two particular passengers, A and B, are sitting on:

a the same side of the minibus

b opposite sides of the minibus.

Answer

a $^3P_2$ ways — A and B both sitting on the driver's side.

$^4P_2$ ways — A and B both not sitting on the driver's side.

$^7P_2$ ways — A and B sitting in any two of the seven seats.

$P(\text{same side}) = P(\text{both on driver's side}) + P(\text{both not on driver's side})$

$$= \frac{^3P_2}{^7P_2} + \frac{^4P_2}{^7P_2}$$

$$= \frac{6}{42} + \frac{12}{42}$$

$$= \frac{3}{7}$$

b $P(\text{opposite sides}) = 1 - P(\text{same side})$

The events 'sitting on the same side' and 'sitting on opposite sides' are complementary.

$$= 1 - \frac{3}{7}$$

$$= \frac{4}{7}$$

Chapter 5: Permutations and combinations

EXERCISE 5G

1. Two children are selected at random from a group of six boys and four girls. Use combinations to find the probability of selecting:

    a two boys
    b two girls
    c one boy and one girl.

2. Three chocolates are selected at random from a box containing 10 milk chocolates and 15 dark chocolates. Find the probability of selecting exactly:

    a two dark chocolates
    b two milk chocolates
    c two dark chocolates or two milk chocolates.

3. Four bananas are randomly selected from a crate of 17 yellow and 23 green bananas. Find the probability that:

    a no green bananas are selected
    b less than half of those selected are green.

4. A curator has 36 paintings and 44 sculptures from which they will randomly select eight items to display in their gallery. Find the probability that the display consists of at least three more paintings than sculptures.

5. Five people are randomly selected from a group of 67 women and 33 men. Find the probability that the selection consists of an odd number of women.

6. In a toolbox there are 25 screwdrivers, 16 drill bits, 38 spanners and 11 chisels. Find the probability that a random selection of four tools contains no chisels.

7. Five clowns each have a red wig and a blue wig, which they are all equally likely to wear at any particular time. Find the probability that, at any particular time:

    a exactly two clowns are wearing red wigs
    b more clowns are wearing blue wigs than red wigs.

8. A gardener has nine rose bushes to plant: three have red flowers and six have yellow flowers. If they plant them in a row in random order, find the probability that:

    a a yellow rose bush is in the middle of the row
    b the three red rose bushes are not separated
    c no two red rose bushes are next to each other.

9. A farmer has 50 animals. They have 24 sheep, of which three are male, and they have 26 cattle, of which 20 are female. A veterinary surgeon wishes to test six randomly selected animals. Find the probability that the selection consists of:

    a equal numbers of cattle and sheep
    b more females than males.

10. a How many distinct arrangements of the letters in the word STATISTICS are there?

    b Find the probability that a randomly selected arrangement begins with:

        i three Ts
        ii three identical letters.

11. Three skirts, four blouses and two jackets are hung in random order on a clothes rail. Find the probability that:

    a the three skirts occupy the middle section of the arrangement
    b the two jackets are not separated.

12. In a group of 180 people, there are 88 males, nine of whom are left-handed, and there are 85 females who are not left-handed. If six people are selected randomly from the group, find the probability that exactly four of them are left-handed or female.

13 A small library holds 1240 books: 312 of the 478 novels (N) have hard covers (H), and there are 440 books that do not have hard covers. Some of this information is shown in the Venn diagram opposite.

a Find the value of a, of b and of c.

b A random selection of 25 of these books is to be donated to a charity group. The charity group hopes that at least 22 of the books will be novels or hard covers. Calculate the probability that the charity group gets what they hope for.

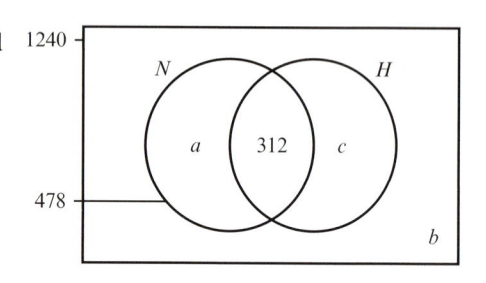

14 A netball team of seven players is to be selected at random from five men and 10 women. Given that at least five women are selected for the team, find the probability that exactly two men are selected.

15 Two items are selected at random from a box that contains some tags and some labels.

Selecting two tags is five times as likely as selecting two labels.

Selecting one tag and one label is six times as likely as selecting two labels.

Find the number of tags and the number of labels in the box.

16 A photograph is to be taken of a pasta dish and n pizzas.

The items are arranged in a line in random order.

Event X is 'the pasta dish is between two pizzas'.

a Investigate the value of $P(X)$ for values of n from 2 to 5.

b Hence, express the value of $\dfrac{P(X')}{P(X)}$ in terms of n. Can you justify your answer for any value of $n \geqslant 2$?

> **WEB LINK**
>
> You will find a range of interesting and challenging probability problems (with hints and solutions) in Module 16 on the NRICH website.

Checklist of learning and understanding

- $n! = n(n-1)(n-2)\ldots \times 3 \times 2 \times 1$, for any integer $n > 0$.

  $0! = 1$

- A key word that points to a permutation is *arranged*.

  A permutation is a way of selecting and arranging objects in a particular order.

- Key words that point to a combination are *chosen* and *selected*.

  A combination is a way of selecting objects in no particular order.

- From n distinct objects, there are:

  $^nP_n = n!$ permutations of all n objects.

  $^nP_r = \dfrac{n!}{(n-r)!}$ permutations of r objects.

  $\dfrac{n!}{p! \times q! \times r! \times \ldots}$ permutations in which there are $p, q, r, \ldots$ of each type.

  $^nC_r = \dfrac{n!}{r!(n-r)!}$ combinations of r objects.

Chapter 5: Permutations and combinations

END-OF-CHAPTER REVIEW EXERCISE 5

1. The word MARMALADE contains four vowels and five consonants. Find the number of possible arrangements of its nine letters if:
   - a there are no restrictions on the order [1]
   - b the arrangement must begin with the four vowels. [2]

2. Five men, four children and two women are asked to stand in a queue at the post office. Find how many ways they can do this if:
   - a the women must be separated [2]
   - b all of the children must be separated from each other. [3]

3. Find the probability that a randomly selected arrangement of all the letters in the word PALLETTE begins and ends with the same letter. [3]

4. Eight-digit mobile phone numbers issued by the Lemon Network all begin with 79.
   - a How many different phone numbers can the network issue? [1]
   - b Find the probability that a randomly selected number issued by this network:
     - i ends with the digits 97 [2]
     - ii reads the same left to right as right to left. [2]

5. There are 12 books on a shelf. Five books are 15 cm tall; four are 20 cm tall and three are 25 cm tall. Find the number of ways that the books can be arranged on the shelf so that none of them is shorter than the book directly to its right. [2]

6. The 11 letters of the word REMEMBRANCE are arranged in a line.
   - i Find the number of different arrangements if there are no restrictions. [1]
   - ii Find the number of different arrangements which start and finish with the letter M. [2]
   - iii Find the number of different arrangements which do not have all 4 vowels (E, E, A, E) next to each other. [3]

   4 letters from the letters of the word REMEMBRANCE are chosen.
   - iv Find the number of different selections which contain no Ms and no Rs and at least 2 Es. [3]

   Cambridge International AS & A Level Mathematics 9709 Paper 62 Q6 November 2013

7. Find how many ways 15 children can be divided into three groups of five if:
   - a there are no restrictions [2]
   - b two of the children are brothers who must be in the same group. [3]

8. An entertainer has been asked to give a performance consisting of four items. They know three songs, five jokes, two juggling tricks and can play one tune on the mandolin. Find how many different ways there are for them to choose the four items if:
   - a there are no restrictions on their performance [1]
   - b they decide not to sing any songs [2]
   - c they are not allowed to tell more than two jokes. [3]

9. From a group of nine people, five are to be chosen at random to serve on a committee. In how many ways can this be done if two particular people refuse to serve on the committee together? [3]

Cambridge International AS & A Level Mathematics: Probability & Statistics 1

10 Twenty teams have entered a tournament. In order to reduce the number of teams to eight, they are put into groups of five and the teams in each group play each other twice. The top two teams in each group progress to the next round. From this point on, teams are paired up, playing each other once with the losing team being eliminated. How many games are played during the whole tournament? [3]

11 A bank provides each account holder with a nine-digit card number that is arranged in three blocks, as shown in the example opposite.

| 4 | 4 | 7 | | 7 | 0 | | 3 | 5 | 3 | 6 |

Find, in index form, the number of card numbers available if:

a there are no restrictions on the digits used [1]

b none of the three blocks can begin with 0 [2]

c the two digits in the second block must not be the same [2]

d the three-, two- and four-digit numbers on the card are even, odd and even, respectively. [3]

12 A basket holds nine flowers: two are pink, three are yellow and four are red. Four of these flowers are chosen at random. Find the probability that at least two of them are red. [4]

13 Find the number of ways in which 11 different pieces of fruit can be shared between three boys so that each boy receives an odd number of pieces of fruit. [5]

14 A bakery wishes to display seven of its 14 types of cake in a row in its shop window. There are six types of sponge cake, five types of cheesecake and three types of fruitcake. Find the number of possible displays that can be made if the bakery places:

a a sponge cake at each end of the row and includes no fruitcakes in the display [2]

b a fruitcake at one end of the row with sponge cakes and cheesecakes placed alternately in the remainder of the row. [4]

15 Five cards, each marked with a different single-digit number from 3 to 7, are randomly placed in a row. Find the probability that the first card in the row is odd and that the three cards in the middle of the row have a sum of 15. [4]

16 Two ordinary fair dice are rolled and the two faces on which they come to rest are hidden by holding the dice together, as shown, and lifted off the table.

The sum of the numbers on the 10 visible faces of the dice is denoted by T.

a Find the number of possible values of T, and find the most likely value of T. [4]

b Calculate the probability that $T \leq 38$. [3]

17 Three ordinary fair dice are rolled. Find the number of ways in which the number rolled with the first die can exceed the sum of the numbers rolled with the second and third dice. Hence, find the probability that this event does not occur in two successive rolls of the three dice. [6]

18 How many even four-digit numbers can be made from the digits 0, 2, 3, 4, 5 and 7, each used at most once, when the first digit cannot be zero? [4]

19 a i Find how many numbers there are between 100 and 999 in which all three digits are different. [3]

  ii Find how many of the numbers in part **i** are odd numbers greater than 700. [4]

  b A bunch of flowers consists of a mixture of roses, tulips and daffodils. Tom orders a bunch of 7 flowers from a shop to give to a friend. There must be at least 2 of each type of flower. The shop has 6 roses, 5 tulips and 4 daffodils, all different from each other. Find the number of different bunches of flowers that are possible. [4]

  Cambridge International AS & A Level Mathematics 9709 Paper 61 Q6 June 2016

20 Three identical cans of cola, 2 identical cans of green tea and 2 identical cans of orange juice are arranged in a row. Calculate the number of arrangements if

  i the first and last cans in the row are the same type of drink, [3]

  ii the 3 cans of cola are all next to each other and the 2 cans of green tea are not next to each other. [5]

  Cambridge International AS & A Level Mathematics 9709 Paper 63 Q4 June 2010

CROSS-TOPIC REVIEW EXERCISE 2

1. Each of the eight players in a chess team plays 12 games against opponents from other teams. The total number of wins, draws and losses for the whole team are denoted by X, Y and Z, respectively.

   a State the value of $X + Y + Z$. [1]

   b Find the least possible value of $Z - X$, given that $Y = 25$. [1]

   c Given that none of the players drew any of their games and that $X - Z = 50$, find the exact mean number of games won by the players. [2]

2. Six books are randomly given to two girls so that each receives at least one book.

   a In how many ways can this be done? [3]

   b Are both girls more likely to receive an odd number or an even number of books? Give a reason for your answer. [2]

3. The 60 members of a ballroom dance society wish to participate in a competition but the coach that has been hired has seats for only 57 people. In how many ways can 57 members be selected if the society's president and vice president must be included? [2]

4. Four discs in two colours and in four sizes are placed in any order on either of two sticks. The following illustration shows one possible arrangement of the four discs.

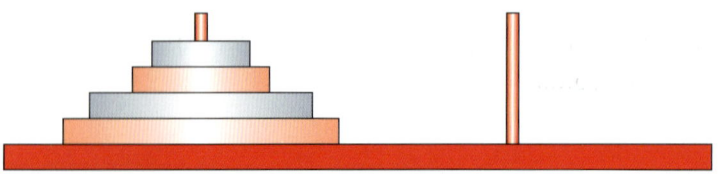

   a Find the number of ways in which the four discs can be arranged so that:

   i they are all on the same stick [2]

   ii there are two discs on each stick. [2]

   b In how many ways can the discs be placed if there are no restrictions? [2]

5. A fair triangular spinner with sides numbered 1, 2 and 3 is spun three times and the numbers that it comes to rest on are written down from left to right to form a three-digit number.

   a How many possible three-digit numbers are there? [1]

   b Find the probability that the three-digit number is:

   i even [1]

   ii odd and greater than 200. [2]

6. A book of poetry contains seven poems, three of which are illustrated. In how many different orders can all the poems be read if no two illustrated poems are read one after the other? [3]

7. Find the number of ways that seven goats and four sheep can sleep in a row if:

   a all the goats must sleep next to each other [2]

   b no two sheep may sleep next to each other. [3]

8. A teacher is looking for 6 pupils to appear in the school play and has decided to select them at random from a group of 11 girls and 13 boys.

   a Find the number of ways in which the teacher can select the 6 pupils. [1]

b Two roles in the play must be played by girls; three roles must be played by boys, but the fool can be played by a girl or by a boy. If the first pupil selected is to play the role of the fool, find the probability that the fool is played by

    i a particular girl [1]

    ii a boy. [1]

c If instead, the pupil who is to play the fool is the last of the six pupils selected, investigate what effect this change in the order of selection has on the probability that the fool is played by:

    i a particular girl [3]

    ii a boy. [3]

9 A radio presenter has enough time at the end of their show to play five songs. She has 13 songs by four groups to choose from: five songs by The Anvils, four by The Braziers, three by The Chisels and one by The Dustbins. Find the number of ways she can choose five songs to play if she decides:

  a that there should be no restrictions [1]

  b to play all three songs by The Chisels [2]

  c to play at least one song by each of the four groups. [4]

10 Students enrolling at an A Level college must select three different subjects to study from the six that are available. One subject must be chosen from each of the option groups A, B and C, as shown in the following table.

Group A	Group B	Group C
Physics	Biology	Mathematics
Chemistry	Physics	Biology
History	Mathematics	Computing

  a One student has chosen to study History and Mathematics. How many subjects do they have to choose from to complete their selection? [1]

  b How many combinations of three subjects are available to a student who enrols at this college? [2]

11 Four ordinary fair dice are arranged in a row. Find the number of ways in which this can be done if the four numbers showing on top of the dice:

  a are all odd [1]

  b have a sum that is less than 7. [3]

12 At company V, 12.5% of the employees have a university degree. At company W, 85% of the employees do not have a university degree. There are 112 employees at company V and 120 employees at company W.

  a One employee is randomly selected. Find the probability that they:

    i work for company V [1]

    ii have a university degree. [2]

  b Five employees from company W are selected at random. Find the probability that none of them has a university degree. [2]

13 One hundred qualified drivers are selected at random. Out of these 100 drivers, of the 40 drivers who wear spectacles, 30 passed their driving test at the first attempt. Altogether, 25 of the drivers did not pass at their first attempt.

  a Show the data given about the drivers in a clearly labelled table or diagram. [3]

  b Did these drivers pass the test at their first attempt independently of whether or not they wear spectacles? Explain your answer. [3]

14 A conference hall has 24 overhead lights. Pairs of lights are operated by switches next to the main entrance, and each switch has three numbered settings: 0 (off), 1 (dim), 2 (bright). Find the number of possible lighting arrangements in the hall if:

   a there are no restrictions [1]

   b two particular pairs of lights must be on setting 2 [1]

   c three lights that are not operated by the same switch and five pairs of lights that are operated by the same switch are not working. [2]

15 Twelve chairs in two colours are arranged, as shown.

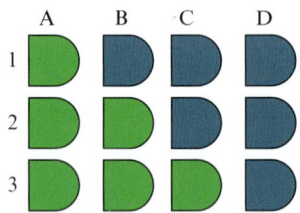

Find in how many ways nine people can sit on these chairs if:

   a the two blue chairs in column C must remain unoccupied [2]

   b all of the green chairs must be occupied [3]

   c more blue chairs than green chairs must be occupied [2]

   d at least one of the chairs in row 2 must remain unoccupied. [5]

16 In a certain country, vehicle registration plates consist of seven characters: a letter, followed by a three-digit number, followed by three letters.

   For example: B 474 PQR

   The first letter cannot be a vowel; the three-digit number cannot begin with 0; and the first of the last three letters cannot be a vowel or any of the letters X, Y or Z.

   a Find the number of registration plates available. [2]

   b Find the probability that a randomly selected registration plate is unassigned, given that there are 48.6 million vehicle owners in the country, and that each owns, on average, 1.183 registered vehicles. [3]

17 Seats for the guests at an awards ceremony are arranged in two rows of eight and ten, divided by an aisle, as shown.

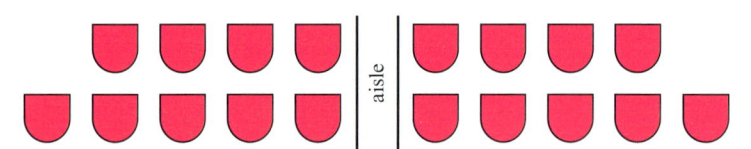

Seats are randomly allocated to 18 guests.

   a Find the probability that two particular guests are allocated seats:

      i on the same side of the aisle [3]

      ii in the same row [3]

      iii on the same side of the aisle and in the same row. [4]

   b Give a reason why the answer to part a iii is not equal to the product of the answers to part a i and part a ii. [1]

Chapter 6
Probability distributions

In this chapter you will learn how to:

- identify and use a discrete random variable
- construct a probability distribution table that relates to a given situation involving a discrete random variable, X, and calculate its expectation, $E(X)$, and its variance, $Var(X)$.

Cambridge International AS & A Level Mathematics: Probability & Statistics 1

PREREQUISITE KNOWLEDGE

Where it comes from	What you should be able to do	Check your skills
IGCSE / O Level Mathematics	Use the fact that $P(A) = 1 - P(A')$.	1 A game can be won (W), lost (L) or drawn (D). Given that $P(W') = 0.46$ and $P(L') = 0.65$, find $P(D)$.
Chapters 4 and 5, sections 4.3, 4.4, 4.5 and 5.4	Distinguish between independent and dependent events, and calculate probabilities accordingly.	2 Two cubes are selected at random from a bag of three red cubes and three blue cubes. Show that the selected cubes are more likely to both be red when the selections are made with replacement than when the selections are made without replacement.

Tools of the trade

Suppose a trading company is planning a new marketing campaign. The campaign will probably go ahead only if the most likely outcome is that sales will increase. However, the company also needs to be aware of worst-case and best-case outcomes, as sales may decrease or decrease dramatically, stay the same or increase dramatically. The company will be able to make informed decisions based on its estimates of the probabilities of these possible outcomes. The likelihood of these outcomes will be based on an analysis of a probability distribution for the changes in sales.

The probability distribution described above acts as a prediction for future sales and the risks involved. Suppose the company is considering entering a new line of business but needs to generate at least $50 000 in revenue before it starts to make a profit. If their probability distribution tells them that there is a 40% chance that revenues will be less than $50 000, then the company knows roughly what level of risk it is facing by entering that new line of business.

6.1 Discrete random variables

A variable is said to be discrete and random if it can take only certain values that occur by chance.

For example, when we buy a carton of six eggs, some may be broken; the number of broken eggs in a carton is a discrete random variable that can take values 0, 1, 2, 3, 4, 5 or 6.

Discrete random variables may arise from independent trials. For example, if we roll four dice then the number of 6s obtained, S, is a discrete random variable with $S \in \{0, 1, 2, 3, 4\}$.

Situations where selections are made without replacement, can also generate discrete random variables. For example, if we randomly select three children from a group of four boys and two girls, the number of boys selected, B, and the number of girls selected, G, are discrete random variables with $B \in \{1, 2, 3\}$ and $G \in \{0, 1, 2\}$.

6.2 Probability distributions

The **probability distribution** of a discrete random variable is a display of all its possible values and their corresponding probabilities. The usual method of display is by tabulation in a probability distribution table. The probability distribution also can be represented in a vertical line graph or in a bar chart.

> **TIP**
> A variable is denoted by an upper-case letter and its possible values by the same lower-case letter. If X can take values of 1, 2 and 3, we write $X \in \{1, 2, 3\}$, where the symbol $\in$ means 'is an element of'.

> **REWIND**
> We learnt how to find probabilities for selections with and without replacement in Chapters 4 and 5, Sections 4.3, 4.4, 4.5 and 5.4.

Consider tossing two fair coins, where we can obtain 0, 1 or 2 heads.

The number of heads obtained in each trial, X, is a discrete random variable and $X \in \{0, 1, 2\}$.

$P(X = 0) = P(\text{tails and tails}) = 0.5 \times 0.5 = 0.25$

$P(X = 1) = P(\text{heads and tails}) + P(\text{tails and heads}) = (0.5 \times 0.5) + (0.5 \times 0.5) = 0.5$

$P(X = 2) = P(\text{heads and heads}) = 0.5 \times 0.5 = 0.25$

The probability distribution for X is displayed in the following table.

x	0	1	2
$P(X = x)$	0.25	0.5	0.25

The probabilities for the possible values of X are equal to the relative frequencies of the values. We would expect 25% of the tosses to produce zero heads; 50% to produce one head and 25% to produce two heads.

WORKED EXAMPLE 6.1

A fair square spinner with sides labelled 1, 2, 3 and 4 is spun twice. The two scores obtained are added together to give the total, X. Draw up the probability distribution table for X.

Answer

2nd spin				
4	5	6	7	8
3	4	5	6	7
2	3	4	5	6
1	2	3	4	5
	1	2	3	4
	1st spin			

The grid shows the 16 equally likely outcomes for the discrete random variable X, where $X \in \{2, 3, 4, 5, 6, 7, 8\}$.

> **TIP**
> $P(X = x)$ is equal to the relative frequency of each particular value of X.

x	2	3	4	5	6	7	8
$P(X = x)$	$\frac{1}{16}$	$\frac{2}{16}$	$\frac{3}{16}$	$\frac{4}{16}$	$\frac{3}{16}$	$\frac{2}{16}$	$\frac{1}{16}$

Sum = 1

The probability distribution for X is shown in the table.

> **TIP**
> Note that $\Sigma P(X = x) = 1$.

WORKED EXAMPLE 6.2

The following table shows the probability distribution for the random variable V.

v	2	3	4	5	6
$P(V = v)$	0.05	c^2	$c + 0.1$	$2c + 0.05$	0.16

Find the value of the constant c and find $P(V > 4)$.

Answer

$$0.05 + c^2 + c + 0.1 + 2c + 0.05 + 0.16 = 1$$
$$c^2 + 3c - 0.64 = 0$$
$$(c - 0.2)(c + 3.2) = 0$$

> We use $\Sigma p = 1$ to form and solve an equation in c.

$c = 0.2$ or $c = -3.2$

The valid solution is $c = 0.2$.

> Note that if $c = -3.2$, then $P(V = 3) = 10.24$, $P(V = 4) = -3.1$ and $P(V = 5) = -6.35$.

$\therefore P(V > 4) = P(V = 5) + P(V = 6)$
$= (2 \times 0.2) + 0.05 + 0.16$
$= 0.61$

> **KEY POINT 6.1**
>
> A probability distribution shows all the possible values of a variable and the sum of the probabilities is $\Sigma p = 1$

> **TIP**
>
> Do check whether the solutions are valid. Remember that a probability cannot be less than 0 or greater than 1.

WORKED EXAMPLE 6.3

There are spaces for three more passengers on a bus, but eight youths, one man and one woman wish to board.

The bus driver decides to select three of these people at random and allow them to board.

Draw up the probability distribution table for Y, the number of youths selected.

Answer

Selections are made without replacement, so we can use combinations to find $P(Y = y)$.

Possible values of Y are 1, 2 and 3

> At least one youth will be selected because there are only two non-youths, who we denote by Y'.

$^{10}C_3$ possible selections.

> Selecting three from 10 people.

$P(Y = 1) = \dfrac{^8C_1 \times {}^2C_2}{^{10}C_3} = \dfrac{1}{15}$

> Selecting one from $8Y$, and two from $2Y'$.

$P(Y = 2) = \dfrac{^8C_2 \times {}^2C_1}{^{10}C_3} = \dfrac{7}{15}$

> Selecting two from $8Y$, and one from $2Y'$.

$P(Y = 3) = \dfrac{^8C_3 \times {}^2C_0}{^{10}C_3} = \dfrac{7}{15}$

> Selecting three from $8Y$, and none from $2Y'$.

y	1	2	3
$P(Y = y)$	$\dfrac{1}{15}$	$\dfrac{7}{15}$	$\dfrac{7}{15}$

> The table shows the probability distribution for Y.

> **TIP**
>
> Always check that $\Sigma p = 1$.

Chapter 6: Probability distributions

EXERCISE 6A

1 The discrete random variable V is such that $V \in \{1, 2, 3\}$. Given that $P(V = 1) = P(V = 2) = 2 \times P(V = 3)$, draw up the probability distribution table for V.

2 The probability distribution for the random variable X is given in the following table.

x	2	3	4	5
$P(X = x)$	p	$2p$	$\frac{1}{2}p$	$3p$

Find the value of p and work out $P(2 < X < 5)$.

3 The probability distribution for the random variable W is given in the following table.

w	3	6	9	12	15
$P(W = w)$	$2k$	k^2	$\frac{k}{2}$	$\frac{4}{5} - 3k$	$\frac{13}{50}$

a Form an equation using k, then solve it.

b Explain why only one of your solutions is valid.

c Find $P(6 \leqslant W < 10)$.

4 The probability that a boy succeeds with each basketball shot is $\frac{7}{9}$. He takes two shots and the discrete random variable S represents the number of successful shots.

Show that $P(S = 0) = \frac{4}{81}$ and draw up the probability distribution table for S.

5 At a garden centre, there is a display of roses: 25 are red, 20 are white, 15 are pink and 5 are orange. Three roses are chosen at random.

a Show that the probability of selecting three red roses is approximately 0.0527.

b Draw up the probability distribution table for the number of red roses selected.

c Find the probability that at least one red rose is selected.

6 Three vehicles from a company's six trucks, five vans, three cars and one motorbike are randomly selected and tested for roadworthiness.

a Show that the probability of selecting three vans is $\frac{2}{91}$.

b Draw up the probability distribution table for the number of vans selected.

c Find the probability that, at most, one van is selected.

7 Five grapes are randomly selected without replacement from a bag containing one red grape and six green grapes.

Name and list the possible values of two discrete random variables in this situation.

State the relationship between the values of your two variables.

8 A pack of five DVDs contains three movies and two documentaries. Three DVDs are selected and the following table shows the probability distribution for M, the number of movies selected.

m	1	2	3
$P(M = m)$	0.3	0.6	0.1

Draw up the probability distribution table for D, the number of documentaries selected.

9 In a particular country, 90% of the population is right-handed and 40% of the population has red hair. Two people are randomly selected from the population. Draw up the probability distribution for X, the number of right-handed, red-haired people selected, and state what assumption must be made in order to do this.

10 A fair 4-sided die, numbered 1, 2, 3 and 5, is rolled twice. The random variable X is the sum of the two numbers on which the die comes to rest.

 a Show that $P(X = 8) = \dfrac{1}{8}$.

 b Draw up the probability distribution table for X, and find $P(X > 6)$.

11 There are eight letters in a post box, and five of them are addressed to Mr Nut. Mr Nut removes four letters at random from the box.

 a Find the probability that none of the selected letters are addressed to Mr Nut.

 b Draw up the probability distribution table for N, the number of selected letters that are addressed to Mr Nut.

 c Describe one significant feature of a vertical line graph or bar chart that could be used to represent the probability distribution for N.

12 A discrete random variable Y is such that $Y \in \{8, 9, 10\}$. Given that $P(Y = y) = ky$, find the value of the constant k.

13 Q is a discrete random variable and $Q \in \{3, 4, 5, 6\}$.

 a Given that $P(Q = q) = cq^2$, find the value of the constant c.

 b Hence, find $P(Q > 4)$.

14 Four books are randomly selected from a box containing 10 novels, 10 reference books and 5 dictionaries. The random variable N represents the number of novels selected.

 a Find the value of $P(N = 2)$, correct to 3 significant figures.

 b Without further calculation, state which of $N = 0$ or $N = 4$ is more likely. Explain the reasons for your answer.

15 In a game, a fair 4-sided spinner with edges labelled 0, 1, 2 and 3 is spun. If a player spins 1, 2 or 3, then that is their score. If a player scores 0, then they spin a fair triangular spinner with edges labelled 0, 1 and 2, and the number they spin is their score. Let the variable X represent a player's score.

 a Show that $P(X = 0) = \dfrac{1}{12}$.

 b Draw up the probability distribution table for X, and find the probability that X is a prime number.

16 A biased coin is tossed three times. The probability distribution for H, the number of heads obtained, is shown in the following table.

h	0	1	2	3
$P(H = h)$	0.512	0.384	0.096	a

   a Find the probability of obtaining a head each time the coin is tossed.

   b Give another discrete random variable that is related to these trials, and calculate the probability that its value is greater than the value of H.

17 Two ordinary fair dice are rolled. A score of 3 points is awarded if exactly one die shows an odd number and there is also a difference of 1 between the two numbers obtained. A player who rolls two even numbers is awarded a score of 2 points, otherwise a player scores 1 point.

   a Draw up the probability distribution table for S, the number of points awarded.

   b Find the probability that a player scores 3 points, given that the sum of the numbers on their two dice is greater than 9.

18 The discrete random variable R is such that $R \in \{1, 3, 5, 7\}$.

   a Given that $P(R = r) = \dfrac{k(r+1)}{r+2}$, find the value of the constant k.

   b Hence, find $P(R \leq 4)$.

EXPLORE 6.1

Consider the probability distribution for X, the number of heads obtained when two fair coins are tossed, which was given in the table presented in the introduction of Section 6.2. Sketch or simply describe the *shape* of a bar chart (or vertical line graph) that can be used to represent this distribution.

In this activity, you will investigate how the shape of the distribution of X is altered when two unfair coins are tossed; that is, when the probability of obtaining heads is $p \neq 0.5$.

Consider the case in which $p = 0.4$ for both coins. Draw a bar chart to represent the probability distribution of X, the number of heads obtained.

Next consider the case in which $p = 0.6$ for both coins, and draw a bar chart to represent the probability distribution of X.

What do you notice about the bar charts for $p = 0.4$ and $p = 0.6$?

Investigate other pairs of probability distributions for which the values of p add up to 1, such as $p = 0.3$ and $p = 0.7$. Make general comments to summarise your results.

Investigate how the value of $P(X = 1)$ changes as p increases from 0 to 1, and then represent this graphically. On the same diagram, show how the values of $P(X = 0)$ and $P(X = 2)$ change as p increases from 0 to 1.

WEB LINK

This can be done manually or using the Coin Flip Simulation on the GeoGebra website.

FAST FORWARD

We will learn how to extend this Explore activity to more than two coins in Chapter 7. We will see how to represent the probability distribution for a continuous random variable in Chapter 8, Section 8.1.

Cambridge International AS & A Level Mathematics: Probability & Statistics 1

> **DID YOU KNOW?**
>
> The simple conjecture of Fermat's Last Theorem, which is that $x^n + y^n = z^n$ has no positive integer solutions for any integer $n > 2$, defeated the greatest mathematicians for 350 years.
>
> The theorem is simple in that it says 'a square can be divided into two squares, but a cube cannot be divided into two cubes, nor a fourth power into two fourth powers, and so on'. Pierre de Fermat himself claimed to have a proof but only wrote in his notebook that 'this margin is too narrow to contain it'!
>
> Fermat's correspondence with the French mathematician, physicist, inventor and philosopher Blaise Pascal helped to develop a very important concept in basic probability that was revolutionary at the time; namely, the idea of equally likely outcomes and expected values.
>
> *Pierre de Fermat, 1607–1665.*
>
> Since his death in 1665, substantial prizes have been offered for a proof, which was finally delivered by Briton Andrew Wiles in 1995.
>
> Wiles' proof used highly advanced 20th century mathematics (i.e. functions of complex numbers in hyperbolic space and the doughnut-shaped solutions of elliptic curves!) that was not available to Fermat.
>
> *Andrew Wiles*

6.3 Expectation and variance of a discrete random variable

Values of a discrete random variable with high probabilities are expected to occur more frequently than values with low probabilities. When a number of trials are carried out, a frequency distribution of values is produced, and this distribution has a mean or *expected* value.

Expectation

The mean of a discrete random variable X is referred to as its expectation, and is written $E(X)$.

Suppose we have a biased spinner with which we can score 0, 1, 2 or 3. The probabilities for these scores, X, are as given in the following table and are also represented in the graph.

x	0	1	2	3
$P(X = x)$	0.1	0.3	0.4	0.2

> **TIP**
>
> We can think of $E(X)$ as being the long-term average value of X over a large number of trials.

However many times it is spun, we expect to score 0 with 10% of the spins; 1 with 30%; 2 with 40% and 3 with 20%.

The expected frequencies of the scores in 1600 trials are shown in the following table.

x	0	1	2	3
Expected frequency (f)	$0.1 \times 1600 = 160$	$0.3 \times 1600 = 480$	$0.4 \times 1600 = 640$	$0.2 \times 1600 = 320$

From this table of expected frequencies, we can calculate the mean (expected) score in 1600 trials.

$$\text{Mean} = E(X) = \frac{\Sigma xf}{\Sigma f} = \frac{(0 \times 160) + (1 \times 480) + (2 \times 640) + (3 \times 320)}{1600} = 1.7$$

We obtain the same value for $E(X)$ if relative frequencies (i.e. probabilities) are used instead of frequencies.

$$\text{Mean} = E(X) = \frac{\Sigma xp}{\Sigma p} = \frac{(0 \times 0.1) + (1 \times 0.3) + (2 \times 0.4) + (3 \times 0.2)}{1} = 1.7$$

> **KEY POINT 6.2**
>
> The expectation of a discrete random variable is $E(X) = \Sigma xp$

> **TIP**
>
> The denominator is $\Sigma p = 1$, so we can omit it from our calculation of $E(X)$.

EXPLORE 6.2

Adam and Priya each have a bag of five cards, numbered 1, 2, 3, 4 and 5. They simultaneously select a card at random from their bag and place it face-up on a table. The numerical difference between the numbers on their cards, X, is recorded, where $X \in \{0, 1, 2, 3, 4\}$. They repeat this 200 times and use their results to draw up a probability distribution table for X.

Adam suggests a new experiment in which the procedure will be the same, except that each of them can choose the card that they place on the table. He says the probability distribution for X will be very different because the cards are not selected at random. Priya disagrees, saying that it will be very similar, or may even be exactly the same.

Do you agree with Adam or Priya? Explain your reasoning.

> **TIP**
>
> An alternative way to write the formula for expectation is $E(X) = \Sigma[x \times P(X = x)]$.

Variance

The variance and standard deviation of a discrete random variable give a measure of the spread of values around the mean, $E(X)$. These measures, like $E(X)$, can be calculated using probabilities in place of frequencies.

If we replace f by p and replace $\bar{x}$ by $E(X)$ in the second of the two formulae for variance, $\frac{\Sigma x^2 f}{\Sigma f} - \bar{x}^2$, we obtain $\frac{\Sigma x^2 p}{\Sigma p} - \{E(X)\}^2$, which simplifies to $\Sigma x^2 p - \{E(X)\}^2$ because $\Sigma p = 1$.

> **REWIND**
>
> We can remember variance from Chapter 3, Section 3.3 as 'mean of the squares minus square of the mean'.

> **KEY POINT 6.3**
>
> The variance of a discrete random variable is $\text{Var}(X) = \Sigma x^2 p - \{E(X)\}^2$.

WORKED EXAMPLE 6.4

The following table shows the probability distribution for X. Find its expectation, variance and standard deviation.

x	0	5	15	20
$P(X = x)$	$\frac{1}{12}$	$\frac{3}{12}$	$\frac{5}{12}$	$\frac{3}{12}$

Answer

$$E(X) = \left(0 \times \frac{1}{12}\right) + \left(5 \times \frac{3}{12}\right) + \left(15 \times \frac{5}{12}\right) + \left(20 \times \frac{3}{12}\right)$$

$$= \frac{1}{12} \times [(0 \times 1) + (5 \times 3) + (15 \times 5) + (20 \times 3)]$$

$$= \frac{1}{12} \times 150$$

$$= 12.5$$

> Substitute values of X and $P(X = x)$ into the formula $E(X) = \Sigma xp$.

> **TIP**
> The working is simpler when all fractions have the same denominator.

$$\text{Var}(X) = \left(0^2 \times \frac{1}{12}\right) + \left(5^2 \times \frac{3}{12}\right) + \left(15^2 \times \frac{5}{12}\right) + \left(20^2 \times \frac{3}{12}\right) - \{12.5\}^2$$

$$= \frac{1}{12} \times [(0 \times 1) + (25 \times 3) + (225 \times 5) + (400 \times 3)] - 156.25$$

$$= \frac{2400}{12} - 156.25$$

$$= 43.75$$

$$SD(X) = \sqrt{43.75} = 6.61, \text{ correct to 3 significant figures.}$$

> Substitute into the formula for $\text{Var}(X)$.

> Take the square root of the variance.

> **TIP**
> Remember to subtract the square of $E(X)$ when calculating variance.

EXERCISE 6B

1 The probability distribution for the random variable X is given in the following table.

x	0	1	2	3
$P(X = x)$	0.10	0.12	0.36	0.42

Calculate $E(X)$ and $\text{Var}(X)$.

2 The probability distribution for the random variable Y is given in the following table.

y	0	1	2	3	4
$P(Y = y)$	0.03	$2p$	0.32	p	0.05

   a Find the value of p.

   b Calculate $E(Y)$ and the standard deviation of Y.

3 The random variable T is such that $T \in \{1, 3, 6, 10\}$. Given that the four possible values of T are equiprobable, find $E(T)$ and $\text{Var}(T)$.

4 The following table shows the probability distribution for the random variable V.

v	1	3	9	m
$P(V = v)$	0.4	0.28	0.14	0.18

Given that $E(V) = 5.38$, find the value of m and calculate $\text{Var}(V)$.

5 R is a random variable such that $R \in \{10, 20, 70, 100\}$. Given that $P(R = r)$ is proportional to r, show that $E(R) = 77$ and find $\text{Var}(R)$.

Chapter 6: Probability distributions

6 The probability distribution for the random variable W is given in the following table.

w	2	7	a	24
$P(W = w)$	0.3	0.3	0.1	0.3

Given that $E(W) = a$, find a and evaluate $Var(W)$.

M 7 The possible outcomes from a business venture are graded from 5 to 1, as shown in the following table.

Grade	5	4	3	2	1
Outcome	High profit	Fair profit	No loss	Small loss	Heavy loss
Probability	0.24	0.33	0.24	0.11	0.08

   a Calculate the expected grade and use it to describe the expected outcome of the venture. Find the standard deviation and explain what it gives a measure of in this case.

   b Investigate the expected outcome and the standard deviation when the grading is reversed (i.e. high profit is graded 1, and so on). Compare these outcomes with those from part **a**.

8 Two ordinary fair dice are rolled. The discrete random variable X is the lowest common multiple of the two numbers rolled.

   a Draw up the probability distribution table for X.

   b Find $E(X)$ and $P[X > E(X)]$.

   c Calculate $Var(X)$.

9 In a game, a player attempts to hit a target by throwing three darts. With each throw, a player has a 30% chance of hitting the target.

   a Draw up the probability distribution table for H, the number of times the target is hit in a game.

   b How many times is the target expected to be hit in 1000 games?

10 Two students are randomly selected from a class of 12 girls and 18 boys.

   a Find the expected number of girls and the expected number of boys.

   b Write the ratio of the expected number of girls to the expected number of boys in simplified form. What do you notice about this ratio?

   c Calculate the variance of the number of girls selected.

11 A sewing basket contains eight reels of cotton: four are green, three are red and one is yellow. Three reels of cotton are randomly selected from the basket.

   a Show that the expected number of yellow cotton reels is 0.375.

   b Find the expected number of red cotton reels.

   c Hence, state the expected number of green cotton reels.

12 A company offers a $1000 cash loan to anyone earning a monthly salary of at least $2000. To secure the loan, the borrower signs a contract with a promise to repay the $1000 plus a fixed fee before 3 months have elapsed. Failure to do this gives the company a legal right to take $1540 from the borrower's next salary before returning any amount that has been repaid.

From past experience, the company predicts that 70% of borrowers succeed in repaying the loan plus the fixed fee before 3 months have elapsed.

Cambridge International AS & A Level Mathematics: Probability & Statistics 1

  a Calculate the fixed fee that ensures the company an expected 40% profit from each $1000 loan.

  b Assuming that the company charges the fee found in part **a**, how would it be possible, without changing the loan conditions, for the company's expected profit from each $1000 loan to be greater than 40%?

13 When a scout group of 8 juniors and 12 seniors meets on a Monday evening, one scout is randomly selected to hoist a flag. Let the variable X represent the number of juniors selected over n consecutive Monday evenings.

  a By drawing up the probability distribution table for X, or otherwise, show that $E(X) = 1.2$ when $n = 3$.

  b Find the number of Monday evenings over which 14 juniors are expected to be selected to hoist the flag.

14 An ordinary fair die is rolled. If the die shows an odd number then S, the score awarded, is equal to that number. If the die shows an even number, then the die is rolled again. If on the second roll it shows an odd number, then that is the score awarded. If the die shows an even number on the second roll, the score awarded is equal to half of that even number.

  a List the possible values of S and draw up a probability distribution table.

  b Find $P[S > E(S)]$.

  c Calculate the exact value of $Var(S)$.

PS 15 A fair 4-sided spinner with sides labelled A, B, B, B is spun four times.

  a Show that there are six equally likely ways to obtain exactly two Bs with the four spins.

  b By drawing up the probability distribution table for X, the number of times the spinner comes to rest on B, find the value of $\dfrac{Var(X)}{E(X)}$.

  c What, in the context of this question, does the value found in part **b** represent?

EXPLORE 6.3

In this activity we will investigate a series of trials in which each can result in one of two possible outcomes.

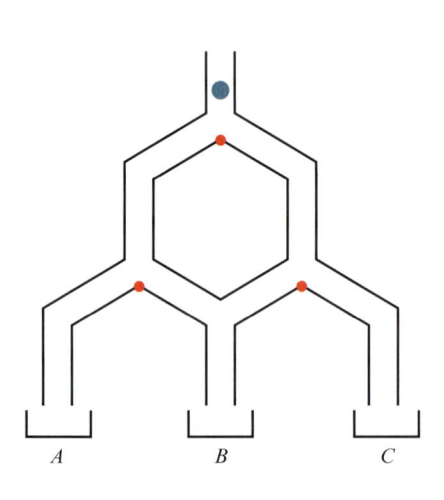

A ball is dropped into the top of the device shown in the diagram.

When a ball hits a nail (which is shown as a red dot), there are two equally likely outcomes: it can fall to the left or it can fall to the right.

Using L and R (to indicate left and right), list all the ways that a ball can fall into each of the cups A, B and C.

Use your lists to tabulate the probabilities of a ball falling into each of the cups. Give all probabilities with denominator 4.

FAST FORWARD

We will study the expectation of two special discrete random variables, and the variance of one of them, in Chapter 7, Sections 7.1 and 7.2.

Chapter 6: Probability distributions

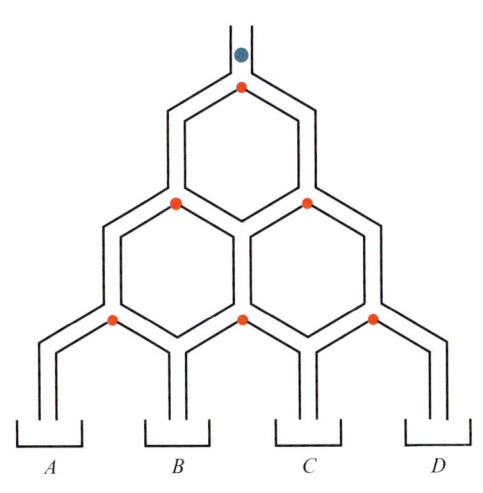

The diagram shows a similar device with four cups labelled A to D.

List all the ways that a ball can fall into each of the cups.

Use your lists to tabulate the probabilities of a ball falling into each of the cups. Give all probabilities with denominator 8.

Can you explain how and why the values $\left(\frac{11}{2}\right)^2$ and $\left(\frac{11}{2}\right)^3$ are connected with the probabilities in your tables?

The next device in the sequence has 10 nails on four rows. Tabulate the probabilities of a ball falling into each of its five cups, A to E.

> **FAST FORWARD**
>
> We will study independent trials that have only two possible outcomes, such as left or right and success or failure, in Chapter 7, Sections 7.1 and 7.2.

Checklist of learning and understanding

- A discrete random variable can take only certain values and those values occur in a certain random manner.
- A probability distribution for a discrete random variable is a display of all its possible values and their corresponding probabilities.
- For the discrete random variable X:

  $\Sigma p = 1$

  $E(X) = \Sigma xp$

  $\text{Var}(X) = \Sigma x^2 p - \{E(X)\}^2$

END-OF-CHAPTER REVIEW EXERCISE 6

1 Find the mean and the variance of the discrete random variable X, whose probability distribution is given in the following table. [3]

x	1	2	3	4
$P(X = x)$	$1-k$	$2-3k$	$3-4k$	$4-6k$

2 The following table shows the probability distribution for the random variable Y.

y	1	10	q	101
$P(Y = y)$	0.2	0.4	0.2	0.2

    a Given that $\text{Var}(Y) = 1385.2$, show that $q^2 - 61q + 624 = 0$ and solve this equation. [4]

    b Find the greatest possible value of $E(Y)$. [2]

3 An investment company has produced the following table, which shows the probabilities of various percentage profits on money invested over a period of 3 years.

Profit (%)	1	5	10	15	20	30	40	45	50
Probability	0.05	0.10	0.50	0.20	0.05	0.04	0.03	0.02	0.01

    a Calculate the expected profit on an investment of $50 000. [3]

    b A woman considers investing $50 000 with the company, but decides that her money is likely to earn more when invested over the same period in a savings account that pays $r\%$ compound interest per annum.

       Calculate, correct to 2 decimal places, the least possible value of r. [3]

4 A chef wishes to decorate each of four cupcakes with one randomly selected sweet. They choose the sweets at random from eight toffees, three chocolates and one jelly. Find the variance of the number of cupcakes that will be decorated with a chocolate sweet. [6]

5 The faces of a biased die are numbered 1, 2, 3, 4, 5 and 6. The random variable X is the score when the die is thrown. The probability distribution table for X is given.

x	1	2	3	4	5	6
$P(X = x)$	p	p	p	p	0.2	0.2

    The die is thrown 3 times. Find the probability that the score is at least 4 on at least 1 of the 3 throws. [5]

    Cambridge international AS & A Level Mathematics 9709 Paper 61 Q2 June 2016 [Adapted]

6 A picnic basket contains five jars: one of marmalade, two of peanut butter and two of jam. A boy removes one jar at random from the basket and then his sister takes two jars, both selected at random.

    a Find the probability that the sister selects her jars from a basket that contains:

        i exactly one jar of jam [1]

        ii exactly two jars of jam. [1]

    b Draw up the probability distribution table for J, the number of jars of jam selected by the sister, and show that $E(J) = 0.8$. [4]

7 Two ordinary fair dice are rolled. The product and the sum of the two numbers obtained are calculated. The score awarded, S, is equal to the absolute (i.e. non-negative) difference between the product and the sum.

For example, if 5 and 3 are rolled, then $S = (5 \times 3) - (5 + 3) = 7$.

    a State the value of S when 1 and 4 are rolled. [1]

    b Draw up a table showing the probability distribution for the 14 possible values of S, and use it to calculate $E(S)$. [5]

8 A fair triangular spinner has sides labelled 0, 1 and 2, and another fair triangular spinner has sides labelled −1, 0 and 1. The score, X, is equal to the sum of the squares of the two numbers on which the spinners come to rest.

    a List the five possible values of X. [1]

    b Draw up the probability distribution table for X. [3]

    c Given that $X < 4$, find the probability that a score of 1 is obtained with at least one of the spinners. [2]

    d Find the exact value of a, such that the standard deviation of X is $\frac{1}{a} \times E(X)$. [3]

9 A discrete random variable X, where $X \in \{2, 3, 4, 5\}$, is such that $P(X = x) = \frac{(b-x)^2}{30}$.

    a Calculate the two possible values of b. [3]

    b Hence, find $P(2 < X < 5)$. [2]

10 Set A consists of the ten digits 0, 0, 0, 0, 0, 0, 2, 2, 2, 4.

Set B consists of the seven digits 0, 0, 0, 0, 2, 2, 2.

One digit is chosen at random from each set. The random variable X is defined as the sum of these two digits.

    i Show that $P(X = 2) = \frac{3}{7}$. [2]

    ii Tabulate the probability distribution of X. [2]

    iii Find $E(X)$ and $Var(X)$. [3]

    iv Given that $X = 2$, find the probability that the digit chosen from set A was 2. [2]

Cambridge International AS & A Level Mathematics 9709 Paper 63 Q5 June 2010

11 The discrete random variable Y is such that $Y \in \{4, 5, 8, 14, 17\}$ and $P(Y = y)$ is directly proportional to $\frac{1}{y+1}$.

Find $P(Y > 4)$. [4]

12 X is a discrete random variable and $X \in \{0, 1, 2, 3\}$. Given that $P(X > 1) = 0.24$, $P(0 < X < 3) = 0.5$ and $P(X = 0 \text{ or } 2) = 0.62$, find $P(X \leq 2 \mid X > 0)$. [5]

13 Four students are to be selected at random from a group that consists of seven boys and x girls. The variables B and G are, respectively, the number of boys selected and the number of girls selected.

    a Given that $P(B = 1) = P(B = 2)$, find the value of x. [3]

    b Given that $G \neq 3$, find the probability that $G = 4$. [3]

14 A box contains 2 green apples and 2 red apples. Apples are taken from the box, one at a time, without replacement. When both red apples have been taken, the process stops. The random variable X is the number of apples which have been taken when the process stops.

  i Show that $P(X = 3) = \frac{1}{3}$. [3]

  ii Draw up the probability distribution table for X. [3]

Another box contains 2 yellow peppers and 5 orange peppers. Three peppers are taken from the box without replacement.

  iii Given that at least 2 of the peppers taken from the box are orange, find the probability that all 3 peppers are orange. [5]

Cambridge International AS & A Level Mathematics 9709 Paper 63 Q7 November 2014

15 In a particular discrete probability distribution the random variable X takes the value $\frac{120}{r}$ with probability $\frac{r}{45}$, where r takes all integer values from 1 to 9 inclusive.

  i Show that $P(X = 40) = \frac{1}{15}$. [2]

  ii Construct the probability distribution table for X. [3]

  iii Which is the modal value of X? [1]

  iv Find the probability that X lies between 18 and 100. [2]

Cambridge International AS & A Level Mathematics 9709 Paper 62 Q5 November 2009

Chapter 7
The binomial and geometric distributions

In this chapter you will learn how to:

- use formulae for probabilities for the binomial and geometric distributions, and recognise practical situations in which these distributions are suitable models
- use formulae for the expectation and variance of the binomial distribution and for the expectation of the geometric distribution.

> **PREREQUISITE KNOWLEDGE**
>
> | Where it comes from | What you should be able to do | Check your skills |
> |---|---|---|
> | Chapter 4 | Calculate expectation in a fixed number of repeated independent trials, given the probability that a particular event occurs. | 1 Two ordinary fair dice are rolled 378 times. How many times can we expect the sum of the two numbers rolled to be greater than 8? |
> | IGCSE / O Level Mathematics

Pure Mathematics 1 | Expand products of algebraic expressions.

Use the expansion of $(a+b)^n$, where n is a positive integer. | 2 Given that $(a+b)^3 = a^3 + 3a^2b + 3ab^2 + b^3$, find the four fractions in the expansion of $\left(\dfrac{1}{4} + \dfrac{3}{4}\right)^3$ and confirm that their sum is equal to 1. |

Two special discrete distributions

Seen in very simple terms, all experiments have just two possible outcomes: success or failure. A business investment can make a profit or a loss; the defendant in a court case is found innocent or guilty; and a batter in a cricket match is either out or not!

In most real-life situations, however, there are many possibilities between success and failure, but taking this yes/no view of the outcomes does allow us to describe certain situations using a **mathematical model**.

Two such situations concern discrete random variables that arise as a result of repeated independent trials, where the probability of success in each trial is constant.

- A **binomial distribution** can be used to model the number of successes in a fixed number of independent trials.
- A **geometric distribution** can be used to model the number of trials up to and including the first success in an infinite number of independent trials.

7.1 The binomial distribution

Consider an experiment in which we roll four ordinary fair dice.

In each independent trial, we can obtain zero, one, two, three or four 6s.

Let the variable R be the number of 6s rolled, then $R \in \{0, 1, 2, 3, 4\}$.

To find the probability distribution for R, we must calculate $P(R = r)$ for all of its possible values.

Using 6 to represent a success and X to represent a failure in each trial, we have:

$P(\text{success}) = P(6) = \dfrac{1}{6}$ and $P(\text{failure}) = P(X) = \dfrac{5}{6}$.

Calculations to find $P(R = r)$ are shown in the following table.

Chapter 7: The binomial and geometric distributions

r	Ways to obtain r successes	No. ways	P(R = r)
0	(XXXX)	$^4C_0 = 1$	$^4C_0\left(\frac{1}{6}\right)^0\left(\frac{5}{6}\right)^4$
1	(6XXX), (X6XX), (XX6X), (XXX6)	$^4C_1 = 4$	$^4C_1\left(\frac{1}{6}\right)^1\left(\frac{5}{6}\right)^3$
2	(66XX), (6X6X), (6XX6), (X66X), (X6X6), (XX66)	$^4C_2 = 6$	$^4C_2\left(\frac{1}{6}\right)^2\left(\frac{5}{6}\right)^2$
3	(666X), (66X6), (6X66), (X666)	$^4C_3 = 4$	$^4C_3\left(\frac{1}{6}\right)^3\left(\frac{5}{6}\right)^1$
4	(6666)	$^4C_4 = 1$	$^4C_4\left(\frac{1}{6}\right)^4\left(\frac{5}{6}\right)^0$

> **TIP**
> Each way of obtaining a particular number of 6s has the same probability.

In the table, we see that $P(R = r) = {^4C_r}\left(\frac{1}{6}\right)^r\left(\frac{5}{6}\right)^{4-r}$. These probabilities are the terms in the binomial expansion of $\left(\frac{1}{6} + \frac{5}{6}\right)^4$.

Using the $\binom{n}{r}$ notation, the five probabilities shown in the previous table are given by

$$P(R = r) = \binom{4}{r}\left(\frac{1}{6}\right)^r\left(\frac{5}{6}\right)^{4-r}.$$

A discrete random variable that meets the following criteria is said to have a binomial distribution and it is defined by its two **parameters**, n and p.

- There are n repeated independent trials.
- n is finite.
- There are just two possible outcomes for each trial (i.e. success or failure).
- The probability of success in each trial, p, is constant.

The random variable is the number of trials that result in a success.

A discrete random variable, X, that has a binomial distribution is denoted by $X \sim B(n, p)$.

> **TIP**
> For work involving binomial expansions, the notation ^nC_r is rarely used nowadays. Your calculator may use this notation but it has mostly been replaced by $\binom{n}{r}$.

> **REWIND**
> We met a series of independent events with just two possible outcomes in the Explore 6.3 activity in Chapter 6, Section 6.3.

KEY POINT 7.1

If $X \sim B(n, p)$ then the probability of r successes is $p_r = \binom{n}{r} p^r (1-p)^{n-r}$.

> **TIP**
> Values of $\binom{n}{r}$ are the coefficients of the terms in a binomial expansion, and give the number of ways of obtaining r successes in n trials.
> $p^r(1-p)^{n-r}$ is the probability for each way of obtaining r successes and $(n-r)$ failures.

For example, if the variable $X \sim B(3, p)$, then $X \in \{0, 1, 2, 3\}$, and we have the following probabilities.

> **TIP**
> Coefficients for power 3 are 1, 3, 3 and 1.

$$P(X=0) = \binom{3}{0} \times p^0 \times q^3 = 1q^3 \qquad P(X=1) = \binom{3}{1} \times p^1 \times q^2 = 3p^1q^2$$

$$P(X=2) = \binom{3}{2} \times p^2 \times q^1 = 3p^2q^1 \qquad P(X=3) = \binom{3}{3} \times p^3 \times q^0 = 1p^3$$

The coefficients in all binomial expansions are symmetric strings of integers. When arranged in rows, they form what has come to be known as Pascal's triangle (named after the French thinker Blaise Pascal). Part of this arrangement is shown in the following diagram, which includes the coefficient for power 0 for completeness.

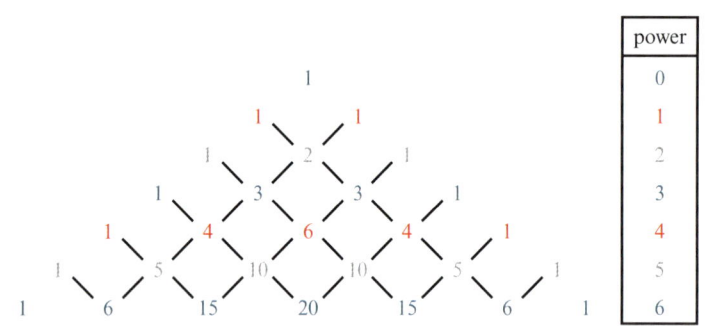

⏮ **REWIND**

We saw in Chapter 5, Section 5.3 that $\binom{n}{r} = {}^nC_r = \dfrac{n!}{r!(n-r)!}$.

power
0
1
2
3
4
5
6

Pascal's triangle rows: 1; 1 1; 1 2 1; 1 3 3 1; 1 4 6 4 1; 1 5 10 10 5 1; 1 6 15 20 15 6 1.

WORKED EXAMPLE 7.1

A regular pentagonal spinner is shown. Find the probability that 10 spins produce exactly three As.

Answer

$P(X=3) = \binom{10}{3} \times 0.4^3 \times 0.6^7$

$= \dfrac{10!}{3! \times 7!} \times 0.4^3 \times 0.6^7$

$= 0.215$ to 3 significant figures.

Let the random variable X be the number of As obtained. We have 10 independent trials with a constant probability of a success, $P(A) = 0.4$. So $X \sim B(10, 0.4)$, and we require three successes and seven failures.

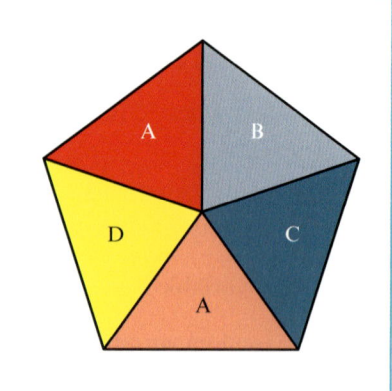

WORKED EXAMPLE 7.2

Given that $X \sim B(8, 0.7)$, find $P(X > 6)$, correct to 3 significant figures.

Answer

$P(X>6) = P(X=7) + P(X=8)$

$= \binom{8}{7} \times 0.7^7 \times 0.3^1 + \binom{8}{8} \times 0.7^8 \times 0.3^0$

$= 0.197650\ldots + 0.057648\ldots$

$= 0.255$

$X \sim B(8, 0.7)$ tells us that $n = 8$, $p = 0.7$, $q = 0.3$, and that $X \in \{0, 1, 2, 3, 4, 5, 6, 7, 8\}$.

💡 **TIP**

Remember that X represents the number of successes, so it can take integer values from 0 to n.

💡 **TIP**

Premature rounding of probabilities in the working may lead to an incorrect final answer. Here, $0.198 + 0.0576 = 0.256$.

Chapter 7: The binomial and geometric distributions

WORKED EXAMPLE 7.3

In a particular country, 85% of the population has rhesus-positive (R+) blood.

Find the probability that fewer than 39 people in a random sample of 40 have rhesus-positive blood.

Answer

$P(X < 39) = 1 - [P(X = 39) + P(X = 40)]$

$= 1 - \left[\binom{40}{39} \times 0.85^{39} \times 0.15^{1} + \binom{40}{40} \times 0.85^{40} \times 0.15^{0} \right]$

$= 1 - [0.010604\ldots + 0.001502\ldots]$

$= 0.988$

Let the random variable X be the number in the sample with R+ blood, then $X \sim B(40, 0.85)$.

REWIND

Recall from Chapter 4, Section 4.1 that $P(A) = 1 - P(A')$.

EXPLORE 7.1

Binomial distributions can be investigated using the Binomial Distribution resource on the GeoGebra website.

We could, for example, check our answer to Worked example 7.3 as follows.

Click on the *distribution* tab and select *binomial* from the pop-up menu at the bottom-left. Select the parameters $n = 40$ and $p = 0.85$, and a bar chart representing the probability distribution will be generated.

To find $P(X < 39)$, enter into the boxes $P(\boxed{0} \leq X \leq \boxed{38})$ and, by tapping the chart, the value for this probability is displayed. (At the right-hand side you will see a list of the probabilities for the 41 possible values of X in this distribution.)

WORKED EXAMPLE 7.4

Given that $X \sim B(n, 0.4)$ and that $P(X = 0) < 0.1$, find the least possible value of n.

Answer

$P(X = 0) = \binom{n}{0} \times 0.4^0 \times 0.6^n = 0.6^n$

We first express $P(X = 0)$ in terms of n.

So we need $0.6^n < 0.1$

$\log 0.6^n < \log 0.1$

$n \log 0.6 < \log 0.1$

$n > \dfrac{\log 0.1}{\log 0.6}$

This leads to an inequality, which we can solve using base 10 logarithms.

$n > 4.50\ldots$

$n = 5$ is the least possible value of n.

TIP

Recall from IGCSE / O Level that if $x > a$, then $-x < -a$. We must reverse the inequality sign when we multiply or divide by a negative number, such as $\log_{10} 0.6$.

TIP

n takes integer values only.

...$0.6^3 = 0.216$; $0.6^4 = 0.1296$; $0.6^5 = 0.07776$.

The least possible value is $n = 5$.

> Alternatively, we can solve $0.6^n < 0.1$ by trial and improvement. We know that n is an integer, so we evaluate $0.6^1, 0.6^2, 0.6^3, \ldots$ up to the first one whose value is less than 0.1.

EXERCISE 7A

1 The variable X has a binomial distribution with $n = 4$ and $p = 0.2$. Find:
   a $P(X = 4)$
   b $P(X = 0)$
   c $P(X = 3)$
   d $P(X = 3 \text{ or } 4)$.

2 Given that $Y \sim B(7, 0.6)$, find:
   a $P(Y = 7)$
   b $P(Y = 5)$
   c $P(Y \neq 4)$
   d $P(3 < Y < 6)$.

3 Given that $W \sim B(9, 0.32)$, find:
   a $P(W = 5)$
   b $P(W \neq 5)$
   c $P(W < 2)$
   d $P(0 < W < 9)$.

4 Given that $V \sim B\left(8, \dfrac{2}{7}\right)$, find:
   a $P(V = 4)$
   b $P(V \geq 7)$
   c $P(V \leq 2)$
   d $P(3 \leq V < 6)$
   e $P(V \text{ is an odd number})$.

5 Find the probability that each of the following events occur.
   a Exactly five heads are obtained when a fair coin is tossed nine times.
   b Exactly two 6s are obtained with 11 rolls of a fair die.

6 A man has five packets and each contains three brown sugar cubes and one white sugar cube. He randomly selects one cube from each packet. Find the probability that he selects exactly one brown sugar cube.

7 A driving test is passed by 70% of people at their first attempt. Find the probability that exactly five out of eight randomly selected people pass at their first attempt.

8 Research shows that the owners of 63% of all saloon cars are male. Find the probability that exactly 20 out of 30 randomly selected saloon cars are owned by:
   a males
   b females.

9 In a particular country, 58% of the adult population is married. Find the probability that exactly 12 out of 20 randomly selected adults are married.

10 A footballer has a 95% chance of scoring each penalty kick that she takes. Find the probability that she:
   a scores from all of her next 10 penalty kicks
   b fails to score from exactly one of her next seven penalty kicks.

11 On average, 13% of all tomato seeds of a particular variety fail to germinate within 10 days of planting. Find the probability that 34 or 35 out of 40 randomly selected seeds succeed in germinating within 10 days of planting.

12 There is a 15% chance of rain on any particular day during the next 14 days. Find the probability that, during the next 14 days, it rains on:

   a exactly 2 days
   b at most 2 days.

13 A factory makes electronic circuit boards and, on average, 0.3% of them have a minor fault. Find the probability that a random sample of 200 circuit boards contains:

   a exactly one with a minor fault
   b fewer than two with a minor fault.

14 There is a 50% chance that a six-year-old child drops an ice cream that they are eating. Ice creams are given to 5 six-year-old children.

   a Find the probability that exactly one ice cream is dropped.
   b 45 six-year-old children are divided into nine groups of five and each child is given an ice cream. Calculate the probability that exactly one of the children in at most one of the groups drops their ice cream.

15 A coin is biased such that heads is three times as likely as tails on each toss. The coin is tossed 12 times. The variables H and T are, respectively, the number of heads and the number of tails obtained. Find the value of $\dfrac{P(H=7)}{P(T=7)}$.

16 Given that $Q \sim B(n, 0.3)$ and that $P(Q=0) > 0.1$, find the greatest possible value of n.

17 The variable $T \sim B(n, 0.96)$ and it is given that $P(T=n) > 0.5$. Find the greatest possible value of n.

18 Given that $R \sim B(n, 0.8)$ and that $P(R > n-1) < 0.006$, find the least possible value of n.

M 19 The number of damaged eggs, D, in cartons of six eggs have been recorded by an inspector at a packing depot. The following table shows the frequency distribution of some of the numbers of damaged eggs in 150 000 boxes.

No. damaged eggs (D)	0	1	2	3	4	5	6
No. cartons (f)	141 393	8396	a	b	0	0	0

The distribution of D is to be modelled by $D \sim B(6, p)$.

   a Estimate a suitable value for p, correct to 4 decimal places.
   b Calculate estimates for the value of a and of b.
   c Calculate an estimate for the least number of additional cartons that would need to be inspected for there to be at least 8400 cartons containing one damaged egg.

M 20 The number of months during the 4-month monsoon season (June to September) in which the total rainfall was greater than 5 metres, R, has been recorded at a location in Meghalaya for the past 32 years, and is shown in the following table.

No. months	0	1	2	3	4
No. years (f)	2	8	12	8	2

The distribution of R is to be modelled by $R \sim B(4, p)$.

   a Find the value of p, and state clearly what this value represents.
   b Give a reason why, in real life, it is unlikely that a binomial distribution could be used to model these data accurately.

21 In a particular country, 90% of both females and males drink tea. Of those who drink tea, 40% of the females and 60% of the males drink it with sugar. Find the probability that in a random selection of two females and two males:

   a all four people drink tea

   b an equal number of females and males drink tea with sugar.

PS 22 It is estimated that 0.5% of all left-handed people and 0.4% of all right-handed people suffer from some form of colour-blindness. A random sample of 200 left-handed and 300 right-handed people is taken. Find the probability that there is exactly one person in the sample that suffers from colour-blindness.

> **DID YOU KNOW?**
>
> Although Pascal's triangle is named after the 17th century French thinker Blaise Pascal, it was known about in China and in Persia as early as the 11th century. The earliest surviving display is of Jia Xian's triangle in a work compiled in 1261 by Yang Hui, as shown in the photo.

> **EXPLORE 7.2**
>
> A frog sits on the bottom-left square of a 5 by 5 grid. In each of the other 24 squares there is a lily pad and four of these have pink flowers growing from them, as shown in the image.
>
> The frog can jump onto an adjacent lily pad but it can only jump northwards (N) or eastwards (E). 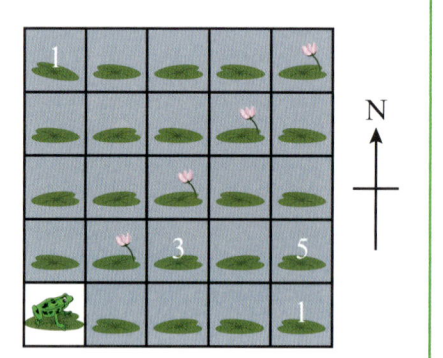
>
> The four numbers on the grid represent the number of different routes the frog can take to get to those particular lily pads. For example, there are three routes to the lily pad with the number 3, and these routes are EEN, ENE and NEE.
>
> Sketch a 5 by 5 grid and write onto it the number of routes to all 24 lily pads. Describe any patterns that you find in the numbers on your grid.
>
> The numbers on the lily pads with pink flowers form a sequence. Can you continue this sequence and find an expression for its nth term?

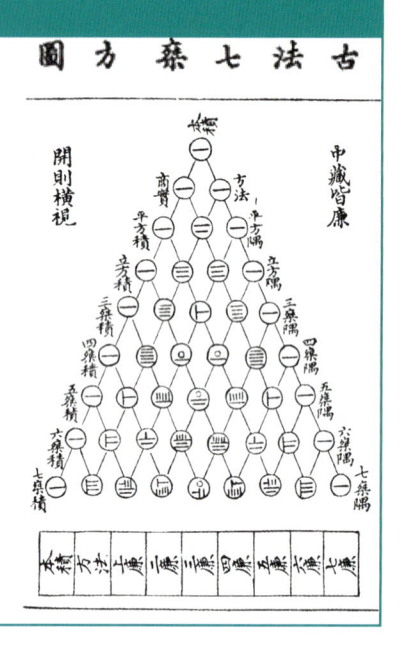

Chapter 7: The binomial and geometric distributions

Expectation and variance of the binomial distribution

Expectation and standard deviation give a measure of central tendency and a measure of variation for the binomial distribution. We can calculate these, along with the variance, from the parameters n and p.

Consider the variable $X \sim B(2, 0.6)$, whose probability distribution is shown in the following table.

x	0	1	2
$P(X = x)$	0.16	0.48	0.36

Applying the formulae for $E(X)$ and $Var(X)$ gives the following results.

$E(X) = \Sigma xp = (0 \times 0.16) + (1 \times 0.48) + (2 \times 0.36) = 1.2$

$Var(X) = \Sigma x^2 p - \{E(X)\}^2 = (0^2 \times 0.16) + (1^2 \times 0.48) + (2^2 \times 0.36) - 1.2^2 = 0.48$

Our experiment consists of $n = 2$ trials with a probability of success $p = 0.6$ in each, so we should not be surprised to find that $E(X) = np = 2 \times 0.6 = 1.2$.

What may be surprising (and a very convenient result), is that the variance of X also can be found from the values of the parameters n and p.

$Var(X) = np(1 - p) = 2 \times 0.6 \times 0.4 = 0.48$

> **REWIND**
>
> We saw in Chapter 6, Section 6.3 that expectation is a variable's long-term average value.

> **REWIND**
>
> We saw in Chapter 4, Section 4.1 that event A is expected to occur $n \times P(A)$ times.

KEY POINT 7.2

The mean and variance of $X \sim B(n, p)$ are given by $\mu = np$ and $\sigma^2 = np(1 - p) = npq$.

> **TIP**
>
> Note that $\mu = E(X)$ and $\sigma^2 = Var(X)$.

WORKED EXAMPLE 7.5

Given that $X \sim B(12, 0.3)$, find the mean, the variance and the standard deviation of X.

Answer

$E(X) = np$
$ = 12 \times 0.3$
$ = 3.6$

$Var(X) = np(1 - p)$
$ = 12 \times 0.3 \times 0.7$
$ = 2.52$

$SD(X) = \sqrt{np(1 - p)}$
$ = \sqrt{2.52}$
$ = 1.59$ to 3 significant figures

> **TIP**
>
> We can also write our answers as $\mu = 3.6$, $\sigma^2 = 2.52$ and $\sigma = 1.59$.

WORKED EXAMPLE 7.6

The random variable $X \sim B(n, p)$. Given that $E(X) = 12$ and $Var(X) = 7.5$, find:

a the value of n and of p

b $P(X = 11)$.

Answer

a $q = \dfrac{7.5}{12} = 0.625$

$p = 1 - q = 0.375$

We use $q = \dfrac{npq}{np} = \dfrac{Var(X)}{E(X)}$ to find p.

$n = \dfrac{12}{0.375} = 32$

$E(X) = np$, so $n = \dfrac{E(X)}{p}$

b $P(X = 11) = \dbinom{32}{11} \times 0.375^{11} \times 0.625^{21}$

$X \sim B(32, 0.375)$

$= 0.138$

EXERCISE 7B

1 Calculate the expectation, variance and standard deviation of each of the following discrete random variables. Give non-exact answers correct to 3 significant figures.

  a $V \sim B(5, 0.2)$
  b $W \sim B(24, 0.55)$
  c $X \sim B(365, 0.18)$
  d $Y \sim B(20, \sqrt{0.5})$

2 Given that $X \sim B(8, 0.25)$, calculate:

  a $E(X)$ and $Var(X)$
  b $P[X = E(X)]$
  c $P[X < E(X)]$.

3 Given that $Y \sim B(11, 0.23)$, calculate:

  a $P(Y \neq 3)$
  b $P[Y < E(Y)]$.

4 Given that $X \sim B(n, p)$, $E(X) = 20$ and $Var(X) = 12$, find:

  a the value of n and of p
  b $P(X = 21)$.

5 Given that $G \sim B(n, p)$, $E(G) = 24\frac{1}{2}$ and $Var(G) = 10\frac{5}{24}$, find:

  a the parameters of the distribution of G
  b $P(G = 20)$.

6 W has a binomial distribution, where $E(W) = 2.7$ and $Var(W) = 0.27$. Find the values of n and p and use them to draw up the probability distribution table for W.

7 Give a reason why a binomial distribution would not be a suitable model for the distribution of X in each of the following situations.

  a X is the height of the tallest person selected when three people are randomly chosen from a group of 10.

  b X is the number of girls selected when two children are chosen at random from a group containing one girl and three boys.

  c X is the number of motorbikes selected when four vehicles are randomly picked from a car park containing 134 cars, 17 buses and nine bicycles.

8 The variable $Q \sim B\left(n, \frac{1}{3}\right)$, and its standard deviation is one-third of its mean. Calculate the non-zero value of n and find $P(5 < Q < 8)$.

9 The random variable $H \sim B(192, p)$, and $E(H)$ is 24 times the standard deviation of H. Calculate the value of p and find the value of k, given that $P(H = 2) = k \times 2^{-379}$.

10 It is estimated that 1.3% of the matches produced at a factory are damaged in some way. A household box contains 462 matches.

   a Calculate the expected number of damaged matches in a household box.

   b Find the variance of the number of damaged matches and the variance of the number of undamaged matches in a household box.

   c Show that approximately 10.4% of the household boxes are expected to contain exactly eight damaged matches.

   d Calculate the probability that at least one from a sample of two household boxes contains exactly eight damaged matches.

11 On average, 8% of the candidates sitting an examination are awarded a merit. Groups of 50 candidates are selected at random.

   a How many candidates in each group are not expected to be awarded a merit?

   b Calculate the variance of the number of merits in the groups of 50.

   c Find the probability that:

      i three, four or five candidates in a group of 50 are awarded merits

      ii three, four or five candidates in both of two groups of 50 are awarded merits.

7.2 The geometric distribution

Consider a situation in which we are attempting to roll a 6 with an ordinary fair die.

How likely are we to get our first 6 on the first roll; on the second roll; on the third roll, and so on?

We can answer these questions using the constant probabilities of success and failure: p and $1 - p$.

P(first 6 on first roll) = $p \to$ a success.

P(first 6 on second roll) = $(1 - p)p \to$ a failure followed by a success.

P(first 6 on third roll) = $(1 - p)^2 p \to$ two failures followed by a success.

The distribution of X, the number of trials up to and including the first success in a series of repeated independent trials, is a discrete random variable whose distribution is called a geometric distribution.

The following table shows the probability that the first success occurs on the rth trial.

r	1	2	3	4	....	n	....
$P(X = r)$	p	$p(1-p)$	$p(1-p)^2$	$p(1-p)^3$	....	$p(1-p)^{n-1}$	....

The values of P($X = r$) in the previous table are the terms of a geometric progression (GP) with first term p and common ratio $1 - p$. The sum of the probabilities is equal to the sum to infinity of the GP.

$$\sum [P(X = r)] = S_\infty = \frac{\text{first term}}{1 - \text{common ratio}} = \frac{p}{1 - (1 - p)} = 1.$$

The sum of the probabilities in a geometric probability distribution is equal to 1.

A discrete random variable, X, is said to have a geometric distribution, and is defined by its parameter p, if it meets the following criteria.

- The repeated trials are independent.
- The repeated trials can be infinite in number.
- There are just two possible outcomes for each trial (i.e. success or failure).
- The probability of success in each trial, p, is constant.

> **REWIND**
>
> We saw in Chapter 6, Section 6.2 that $\Sigma p = 1$ for a probability distribution. You will also have seen geometric progressions and geometric series in Pure Mathematics 1, Chapter 6.

KEY POINT 7.3

A random variable X that has a geometric distribution is denoted by $X \sim \text{Geo}(p)$, and the probability that the first success occurs on the rth trial is

$p_r = p(1 - p)^{r-1}$ for $r = 1, 2, 3, \ldots$

> **TIP**
>
> An alternative form of this formula, $P(X = r) = q^{r-1} \times p$, where $p = 1 - q$, reminds us that the $r - 1$ failures occur before the first success.

The binomial and geometric distributions arise in very similar situations. The significant difference is that the number of trials in a binomial distribution is fixed from the start and the number of successes are counted, whereas, in a geometric distribution, trials are repeated as many times as necessary until the first success occurs.

For $X \sim B(n, p)$, there are $\binom{n}{r}$ ways to obtain r successes.

For $X \sim \text{Geo}(p)$, there is only one way to obtain the first success on the rth trial, and that is when there are $r - 1$ failures followed by a success.

> **REWIND**
>
> Recall from Section 7.1 that
> $\binom{n}{r} = {}^nC_r = \frac{n!}{r!(n-r)!}$.

WORKED EXAMPLE 7.7

Repeated independent trials are carried out in which the probability of success in each trial is 0.66.

Correct to 3 significant figures, find the probability that the first success occurs:

a on the third trial

b on or before the second trial

c after the third trial.

Answer

a $P(X = 3) = p(1-p)^2$
   $= 0.66 \times 0.34^2$
   $= 0.0763$

> Let X represent the number of trials up to and including the first success, then $X \sim \text{Geo}(0.66)$, where $p = 0.66$ and $1 - p = 0.34$.

b $P(X \leq 2) = P(X = 1) + P(X = 2)$
   $= p + p(1-p)$
   $= 0.884$

c $P(X > 3) = 1 - P(X \leq 3)$
   $= 1 - [P(X = 1) + P(X = 2) + P(X = 3)]$
   $= 1 - [p + p(1-p) + p(1-p)^2]$
   $= 0.0393$

Probabilities that involve inequalities can be found by summation for small values of r, as in parts **b** and **c** of Worked example 7.7. However, for larger values of r, the following results will be useful.

$P(X \leq r) = P(\text{success on one of the first } r \text{ trials}) = 1 - P(\text{failure on the first } r \text{ trials})$

$P(X > r) = P(\text{first success after the } r\text{th trial}) = P(\text{failure on the first } r \text{ trials})$

These two results are written in terms of q in Key point 7.4.

> **KEY POINT 7.4**
>
> When $X \sim \text{Geo}(p)$ and $q = 1 - p$, then
> - $P(X \leq r) = 1 - q^r$
> - $P(X > r) = q^r$

WORKED EXAMPLE 7.8

In a particular country, 18% of adults wear contact lenses. Adults are randomly selected and interviewed one at a time. Find the probability that the first adult who wears contact lenses is:

a one of the first 15 interviewed

b not one of the first nine interviewed.

Answer

a $P(X \leq 15) = 1 - q^{15}$
   $= 1 - 0.82^{15}$
   $= 0.949$

> Let X represent the number of adults interviewed up to and including the first one who wears contact lenses, then $X \sim \text{Geo}(0.18)$ and $q = 1 - 0.18 = 0.82$.

b $P(X > 9) = q^9$
   $= 0.82^9$
   $= 0.168$

WORKED EXAMPLE 7.9

A coin is biased such that the probability of obtaining heads with each toss is equal to $\frac{5}{11}$. The coin is tossed until the first head is obtained. Find the probability that the coin is tossed:

  a at least six times

  b fewer than eight times.

Answer

a $P(X \geq 6) = P(X > 5)$
  $= q^5$
  $= \left(\frac{6}{11}\right)^5$
  $= 0.0483$

Let X represent the number of times the coin is tossed up to and including the first heads, then $X \sim \text{Geo}\left(\frac{5}{11}\right)$ and $q = \frac{6}{11}$.

TIP
'At least six times' has the same meaning as 'more than five times'.

b $P(X < 8) = P(X \leq 7)$
  $= 1 - q^7$
  $= 1 - \left(\frac{6}{11}\right)^7$
  $= 0.986$

TIP
'Fewer than eight times' has the same meaning as 'seven or fewer times'.

EXERCISE 7C

1 Given the discrete random variable $X \sim \text{Geo}(0.2)$, find:

  a $P(X = 7)$ b $P(X \neq 5)$ c $P(X > 4)$.

2 Given that $T \sim \text{Geo}(0.32)$, find:

  a $P(T = 3)$ b $P(T \leq 6)$ c $P(T > 7)$.

3 The probability that Mike is shown a yellow card in any football match that he plays is $\frac{1}{2}$. Find the probability that Mike is next shown a yellow card:

  a in the third match that he plays   b before the fourth match that he plays.

4 On average, Diya concedes one penalty in every six hockey matches that she plays. Find the probability that Diya next concedes a penalty:

  a in the eighth match that she plays   b after the fourth match that she plays.

5 The sides of a fair 5-sided spinner are marked 1, 1, 2, 3 and 4. It is spun until the first score of 1 is obtained. Find the probability that it is spun:

  a exactly twice   b at most five times   c at least eight times.

Chapter 7: The binomial and geometric distributions

6   It is known that 80% of the customers at a DIY store own a discount card. Customers queuing at a checkout are asked if they own a discount card.

   a   Find the probability that the first customer who owns a discount card is:

      i   the third customer asked
      ii not one of the first four customers asked.

   b   Given that 10% of the customers with discount cards forget to bring them to the store, find the probability that the first customer who owns a discount card and remembered to bring it to the store is the second customer asked.

7   In a manufacturing process, the probability that an item is faulty is 0.07. Items from those produced are selected at random and tested.

   a   Find the probability that the first faulty item is:

      i   the 12th item tested
      ii not one of the first 10 items tested
      iii one of the first eight items tested.

   b   What assumptions have you made about the occurrence of faults in the items so that you can calculate the probabilities in part **a**?

8   Two independent random variables are $X \sim \text{Geo}(0.3)$ and $Y \sim \text{Geo}(0.7)$. Find:

   a $P(X = 2)$ b $P(Y = 2)$ c $P(X = 1 \text{ and } Y = 1)$.

9   On average, 14% of the vehicles being driven along a stretch of road are heavy goods vehicles (HGVs). A girl stands on a footbridge above the road and counts the number of vehicles, up to and including the first HGV that passes. Find the probability that she counts:

   a   at most three vehicles   b   at least five vehicles.

10 The probability that a woman can connect to her home Wi-Fi at each attempt is 0.44. Find the probability that she fails to connect until her fifth attempt.

11 Decide whether or not it would be appropriate to model the distribution of X by a geometric distribution in the following situations. In those cases for which it is not appropriate, give a reason.

   a   A bag contains two red sweets and many more green sweets. A child selects a sweet at random and eats it, selects another and eats it, and so on. X is the number of sweets selected and eaten, up to and including the first red sweet.

   b   A monkey sits in front of a laptop with a blank word processing document on its screen. X is the number of keys pressed by the monkey, up to and including the first key pressed that completes a row of three letters that form a meaningful three-letter word.

   c X is the number of times that a grain of rice is dropped from a height of 2 metres onto a chessboard, up to and including the first time that it comes to rest on a white square.

   d X is the number of races in which an athlete competes during a year, up to and including the first race that he wins.

12 The random variable T has a geometric distribution and it is given that $\dfrac{P(T = 2)}{P(T = 5)} = 15.625$. Find $P(T = 3)$.

PS 13 $X \sim \text{Geo}(p)$ and $P(X = 2) = 0.2464$. Given that $p < 0.5$, find $P(X > 3)$.

PS 14 Given that $X \sim \text{Geo}(p)$ and that $P(X \leq 4) = \dfrac{2385}{2401}$, find $P(1 \leq X < 4)$.

PS 15 Two ordinary fair dice are rolled simultaneously. Find the probability of obtaining:

　　a　the first double on the fourth roll

　　b　the first pair of numbers with a sum of more than 10 before the 10th roll.

PS 16 $X \sim \text{Geo}(0.24)$ and $Y \sim \text{Geo}(0.25)$ are two independent random variables. Find the probability that $X + Y = 4$.

Mode of the geometric distribution

All geometric distributions have two features in common. These are clear to see when bar charts or vertical line graphs are used to represent values of $P(X = r)$ for different values of the parameter p. You can do this manually or using a graphing tool such as GeoGebra.

The first common feature is that $P(X = 1)$ has the greatest probability in all geometric distributions. This means that the most likely value of X is 1, so the first success is most likely to occur on the first trial. Secondly, the value of $P(X = r)$ decreases as r increases. This is because the common ratio between the probabilities ($q = 1 - p$) is less than 1:
$$p > p(1-p) > p(1-p)^2 > p(1-p)^3 > p(1-p)^4 > \ldots.$$

The following table shows some probabilities for the distributions $X \sim \text{Geo}(0.2)$ and $X \sim \text{Geo}(0.7)$. In both distributions, we can see that probabilities decrease as the value of X increases.

KEY POINT 7.5

The mode of all geometric distributions is 1.

	P(X = 1)	P(X = 2)	P(X = 3)	P(X = 4)	P(X = 5)
Geo(0.2)	0.2	0.16	0.128	0.1024	0.08192
Geo(0.7)	0.7	0.21	0.063	0.0189	0.00567

Expectation of the geometric distribution

Recall that the expectation or mean of a discrete random variable is its long-term average, which is given by $E(X) = \mu = \Sigma x p_x$.

Applying this to the geometric distribution Geo(p), it turns out we find that the mean is equal to $\dfrac{1}{p}$, the reciprocal of p.

KEY POINT 7.6

If $X \sim \text{Geo}(p)$ then $\mu = \dfrac{1}{p}$.

EXPLORE 7.3

Using algebra, we can prove that the mean of the geometric distribution is equal to $\dfrac{1}{p}$.

For $X \sim \text{Geo}(p)$, we have $X \in \{1, 2, 3, 4, \ldots\}$ and $p_x = \{p, pq, pq^2, pq^3, \ldots\}$.

Step 1 of the proof is to form an equation that expresses μ in terms of p and q.

To do this we use $\mu = \Sigma x p_x$.

There are three more steps required to complete the proof, which you might like to try without any further assistance. However, some guidance is given below if needed.

Step 2: Multiply the equation obtained in step 1 throughout by q to obtain a second equation.

REWIND

We studied the expectation of a discrete random variable in Chapter 6, Section 6.3.

Step 3: Subtract one equation from the other.

Step 4: If you have successfully managed steps 1, 2 and 3, you should need no help completing the proof!

WORKED EXAMPLE 7.10

One in four boxes of Zingo breakfast cereal contains a free toy. Let the random variable X be the number of boxes that a child opens, up to and including the one in which they find their first toy.

  a Find the mode and the expectation of X.

  b Interpret the two values found in part **a** in the context of this question.

Answer

  a The mode of X is 1.

$$E(X) = \frac{1}{p} = \left(\frac{1}{4}\right)^{-1} = 4 \quad \cdots \text{The variable is } X \sim \text{Geo}\left(\frac{1}{4}\right).$$

  b A child is most likely to find their first toy in the first box they open but, on average, a child will find their first toy in the fourth box that they open.

> **TIP**
> An answer written 'in context' must refer to a specific situation; in this case, the situation described in the question.

WORKED EXAMPLE 7.11

The variable X follows a geometric distribution. Given that $E(X) = 3\frac{1}{2}$, find $P(X > 6)$.

Answer

$E(X) = \dfrac{1}{p} = \dfrac{7}{2}$, so $p = \dfrac{2}{7}$ We find the parameter p and then we find q.

$q = 1 - \dfrac{2}{7} = \dfrac{5}{7}$

$P(X > 6) = q^6$ We use $P(X > r) = q^r$ from Key point 7.4.

$= \left(\dfrac{5}{7}\right)^6$

$= 0.133$

WORKED EXAMPLE 7.12

Given that $X \sim \text{Geo}(p)$ and that $P(X \leq 3) = \dfrac{819}{1331}$, find:

a $P(X > 3)$

b $P(1 < X \leq 3)$.

Answer

a $P(X > 3) = 1 - P(X \leq 3)$

$= 1 - \dfrac{819}{1331}$

$= \dfrac{512}{1331}$

b $1 - q^3 = P(X \leq 3)$ ············ We use $1 - q^r = P(X \leq r)$ to find q and p.

$q^3 = 1 - \dfrac{819}{1331}$

$q = \sqrt[3]{1 - \dfrac{819}{1331}}$

$q = \dfrac{8}{11}$ and $p = \dfrac{3}{11}$

$P(1 < X \leq 3) = P(X = 2) + P(X = 3)$

$= pq + pq^2$

$= \dfrac{456}{1331}$ or 0.343

TIP

Alternatively, we can use

$P(1 < X \leq 3)$
$= P(X \leq 3) - P(X = 1)$
$= \dfrac{819}{1331} - \dfrac{3}{11}$
$= \dfrac{456}{1331}$

EXERCISE 7D

1 Given that $X \sim \text{Geo}(0.36)$, find the exact value of $E(X)$.

2 The random variable Y follows a geometric distribution. Given that $P(Y = 1) = 0.2$, find $E(Y)$.

3 Given that $S \sim \text{Geo}(p)$ and that $E(S) = 4\dfrac{1}{2}$, find $P(S = 2)$.

4 Let T be the number of times that a fair coin is tossed, up to and including the toss on which the first tail is obtained. Find the mode and the mean of T.

5 Let X be the number of times an ordinary fair die is rolled, up to and including the roll on which the first 6 is obtained. Find $E(X)$ and evaluate $P[X > E(X)]$.

6 A biased 4-sided die is numbered 1, 3, 5 and 7. The probability of obtaining each score is proportional to that score.

   a Find the expected number of times that the die will be rolled, up to and including the roll on which the first non-prime number is obtained.

   b Find the probability that the first prime number is obtained on the third roll of the die.

Chapter 7: The binomial and geometric distributions

7   Sylvie and Thierry are members of a choir. The probabilities that they can sing a perfect high C note on each attempt are $\frac{4}{7}$ and $\frac{5}{8}$, respectively.

   a   Who is expected to fail fewer times before singing a high C note for the first time?

   b   Find the probability that both Sylvie and Thierry succeed in singing a high C note on their second attempts.

PS 8   A standard deck of 52 playing cards has an equal number of hearts, spades, clubs and diamonds. A deck is shuffled and a card is randomly selected. Let X be the number of cards selected, up to and including the first diamond.

   a   Given that X follows a geometric distribution, describe the way in which the cards are selected, and give the reason for your answer.

   b   Find the probability that:

   i X is equal to $E(X)$

   ii   neither of the first two cards selected is a heart and the first diamond is the third card selected.

PS 9   A study reports that a particular gene in 0.2% of all people is defective. X is the number of randomly selected people, up to and including the first person that has this defective gene. Given that $P(X \leq b) > 0.865$, find $E(X)$ and find the smallest possible value of b.

PS 10   Anouar and Zane play a game in which they take turns at tossing a fair coin. The first person to toss heads is the winner. Anouar tosses the coin first, and the probability that he wins the game is $0.5^1 + 0.5^3 + 0.5^5 + 0.5^7 + \ldots$.

   a   Describe the sequence of results represented by the value 0.5^5 in this series.

   b   Find, in a similar form, the probability that Zane wins the game.

   c   Find the probability that Anouar wins the game.

EXPLORE 7.4

In a game for two people that cannot be drawn, you are the stronger player with a 60% chance of winning each game.

The probability distributions for the number of games won by you and those won by your opponent when a single game is played, X and Y, are shown.

x	0	1
$P(X = x)$	0.40	0.60

y	0	1
$P(Y = y)$	0.60	0.40

Investigate the probability distributions for X and Y in a best-of-three contest, where the first player to win two games wins the contest.

Who gains the advantage as the number of games played in a contest increases? What evidence do you have to support your answer?

PS   How likely are you to win a best-of-five contest?

Checklist of learning and understanding

- A binomial distribution can be used to model the number of successes in a series of n repeated independent trials where the probability of success on each trial, p, is constant.
  - If $X \sim B(n, p)$ then $p_r = \binom{n}{r} p^r (1-p)^{n-r}$.
  - $E(X) = \mu = np$
  - $Var(X) = \sigma^2 = np(1-p) = npq$, where $q = 1 - p$.

- A geometric distribution can be used to model the number of trials up to and including the first success in a series of repeated independent trials where the probability of success on each trial, p, is constant.
  - If $X \sim Geo(p)$ then $p_r = p(1-p)^{r-1}$ for $r = 1, 2, 3, \ldots$
  - $E(X) = \mu = \dfrac{1}{p}$
  - $P(X \leqslant r) = 1 - q^r$ and $P(X > r) = q^r$, where $q = 1 - p$.
  - The mode of all geometric distributions is 1.

Chapter 7: The binomial and geometric distributions

END-OF-CHAPTER REVIEW EXERCISE 7

1 Given that $X \sim B\left(n, \dfrac{1}{n}\right)$ find an expression for P($X = 1$) in terms of n. [2]

2 A family has booked a long holiday in Skragness, where the probability of rain on any particular day is 0.3. Find the probability that:

   a the first day of rain is on the third day of their holiday [1]

   b it does not rain for the first 2 weeks of their holiday. [2]

3 One plastic robot is given away free inside each packet of a certain brand of biscuits. There are four colours of plastic robot (red, yellow, blue and green) and each colour is equally likely to occur. Nick buys some packets of these biscuits. Find the probability that

   i he gets a green robot on opening his first packet, [1]

   ii he gets his first green robot on opening his fifth packet. [2]

   Nick's friend Amos is also collecting robots.

   iii Find the probability that the first four packets Amos opens all contain different coloured robots. [3]

Cambridge International AS & A Level Mathematics 9709 Paper 62 Q3 November 2015

4 Weiqi has two fair triangular spinners. The sides of one spinner are labelled 1, 2, 3, and the sides of the other are labelled 2, 3, 4. Weiqi spins them simultaneously and notes the two numbers on which they come to rest.

   a Find the probability that these two numbers differ by 1. [2]

   b Weiqi spins both spinners simultaneously on 15 occasions. Find the probability that the numbers on which they come to rest do not differ by 1 on exactly eight or nine of the 15 occasions. [3]

5 A computer generates random numbers using any of the digits 0, 1, 2, 3, 4, 5, 6, 7, 8, 9. The numbers appear on the screen in blocks of five digits, such as 50119 26317 40068 .... Find the probability that:

   a there are no 7s in the first block [1]

   b the first zero appears in the first block [1]

   c the first 9 appears in the second block. [2]

6 Four ordinary fair dice are rolled.

   a In how many ways can the four numbers obtained have a sum of 22? [2]

   b Find the probability that the four numbers obtained have a sum of 22. [2]

   c The four dice are rolled on eight occasions. Find the probability that the four numbers obtained have a sum of 22 on at least two of these occasions. [3]

7 When a certain driver parks their car in the evenings, they are equally likely to remember or to forget to switch off the headlights. Giving your answers in their simplest index form, find the probability that on the next 16 occasions that they park their car in the evening, they forget to switch off the headlights:

   a 14 more times than they remember to switch them off [2]

   b at least 12 more times than they remember to switch them off. [3]

8 Gina has been observing students at a university. Her data indicate that 60% of the males and 70% of the females are wearing earphones at any given time. She decides to interview randomly selected students and to interview males and females alternately.

a Use Gina's observation data to find the probability that the first person not wearing earphones is the third male interviewed, given that she first interviews:

 i a male   [2]

 ii a female   [2]

 iii a male who is wearing earphones.   [2]

b State any assumptions made about the wearing of earphones in your calculations for part **a**.   [1]

9 In Restaurant Bijoux 13% of customers rated the food as 'poor', 22% of customers rated the food as 'satisfactory' and 65% rated it as 'good'. A random sample of 12 customers who went for a meal at Restaurant Bijoux was taken.

 i Find the probability that more than 2 and fewer than 12 of them rated the food as 'good'.   [3]

On a separate occasion, a random sample of n customers who went for a meal at the restaurant was taken.

 ii Find the smallest value of n for which the probability that at least 1 person will rate the food as 'poor' is greater than 0.95.   [3]

Cambridge International AS & A Level Mathematics 9709 Paper 62 Q3 June 2012

10 A biased coin is four times as likely to land heads up compared with tails up. The coin is tossed k times so that the probability that it lands tails up on at least one occasion is greater than 99%. Find the least possible value of k.   [4]

11 Given that $X \sim B(n, 0.4)$ and that $P(X = 1) = k \times P(X = n - 1)$, express the constant k in terms of n, and find the smallest value of n for which $k > 25$.   [5]

12 A book publisher has noted that, on average, one page in eight contains at least one spelling error, one page in five contains at least one punctuation error, and that these errors occur independently and at random. The publisher checks 480 randomly selected pages from various books for errors.

 a How many pages are expected to contain at least one of both types of error?   [2]

 b Find the probability that:

  i the first spelling error occurs after the 10th page   [2]

  ii the first punctuation error occurs before the 10th page   [2]

  iii the 10th page is the first to contain both types of error.   [2]

13 Robert uses his calculator to generate 5 random integers between 1 and 9 inclusive.

 i Find the probability that at least 2 of the 5 integers are less than or equal to 4.   [3]

Robert now generates n random integers between 1 and 9 inclusive. The random variable X is the number of these n integers which are less than or equal to a certain integer k between 1 and 9 inclusive. It is given that the mean of X is 96 and the variance of X is 32.

 ii Find the values of n and k.   [4]

Cambridge International AS & A Level Mathematics 9709 Paper 62 Q4 June 2013

14 Anna, Bel and Chai take turns, in that order, at rolling an ordinary fair die. The first person to roll a 6 wins the game.

Find the ratio P(Anna wins) : P(Bel wins) : P(Chai wins), giving your answer in its simplest form.   [7]

Chapter 8
The normal distribution

In this chapter you will learn how to:

- sketch normal curves to illustrate distributions or probabilities
- use a normal distribution to model a continuous random variable and use normal distribution tables
- solve problems concerning a normally distributed variable
- recognise conditions under which the normal distribution can be used as an approximation to the binomial distribution, and use this approximation, with a continuity correction, in solving problems.

Cambridge International AS & A Level Mathematics: Probability & Statistics 1

PREREQUISITE KNOWLEDGE

Where it comes from	What you should be able to do	Check your skills
Chapter 7	Find and calculate with the expectation and variance of a binomial distribution.	1 Given $X \sim \text{B}(45, 0.52)$, find $\text{E}(X)$ and $\text{Var}(X)$. 2 Given that X follows a binomial distribution with $\text{E}(X) = 11.2$ and $\text{Var}(X) = 7.28$, find the parameters of the distribution of X.

Why are errors quite normal?

If you study any of the sciences, you will be required at some time to measure a quantity as part of an experiment. That quantity could be a measurement of time, mass, distance, volume and so on. Whatever it is, any measurement you make of a continuous quantity such as these will be subject to error. The very nature of continuous quantities means that they cannot be measured precisely and, no matter how hard we try, inaccuracy is also likely because our tools lack perfect calibration and we, as human beings, add in a certain amount of unreliability.

However, small errors are more likely than large errors and our measurements are usually just as likely to be underestimates as overestimates. When repeated measurements are taken, errors are likely to cancel each other out, so the average error is close to zero and the average of the measurements is virtually error-free.

This chapter serves as an introduction to the idea of a continuous random variable and the method used to display its probability distribution. We will later focus our attention on one particular type of continuous random variable, namely a normal random variable.

The **normal distribution** was discovered in the late 18th century by the German mathematician Carl Friedrich Gauss through research into the measurement errors made in astronomical observations. Some key properties of the normal distribution are that values close to the average are most likely; the further values are from the average, the less likely they are to occur, and the distribution is symmetrical about the average.

8.1 Continuous random variables

A continuous random variable is a quantity that is liable to change and whose infinite number of possible values are the numerical outcomes of a random phenomenon. Examples include the amount of sugar in an orange, the time required to run a marathon, measurements of height and temperature and so on. A continuous random variable is not defined for specific values. Instead, it is defined over an interval of values.

Consider the mass of an apple, denoted by X grams. Within the range of possible masses, X can take any value, such as 111.2233..., or 137.8642..., or 145.2897..., or .... The probability that X takes a particular value is necessarily equal to 0, since the number of values that it can take is infinite. However, there will be a countable number of values in any chosen interval, such as $130 \leq X < 140$, so a probability for each and every interval can be found.

The probability distribution of a discrete random variable shows its specific values and their probabilities, as we saw in Chapters 6 and 7.

The probability distribution of a continuous random variable shows its range of values and the probabilities for intervals within that range.

- When X is a discrete random variable, we can represent $P(X = r)$.
- When X is a continuous random variable, we can represent $P(a \leq X \leq b)$.

Before looking at probability distributions for continuous random variables in detail, we will consider how we can represent the probability distribution of a set of collected or observed continuous data.

Representation of a probability distribution

A set of continuous data can be illustrated in a histogram, where column areas are proportional to frequencies. To illustrate the probability distribution of a set of data, we draw a graph that is based on the shape of a histogram, as we now describe.

If we change the frequency density values on the vertical axis to relative frequency density values (relative frequency density = relative frequency ÷ class width) then column areas will represent relative frequencies, which are estimates of probabilities. The vertical axis of the diagram can now be labeled 'probability density'.

For equal-width class intervals, the process described above has no effect on the 'shape' of the diagram. The result is that the total area of the columns changes from 'Σf' to 1, which is the sum of the probabilities of all the possible values.

So we can draw a curved graph over the columns of an equal-width interval histogram (preferably one displaying large amounts of data with many classes) to model the probability distribution of a set of continuous data.

In the case of a random variable, such a curved graph represents a function, $y = f(x)$, and is called a **probability density function**, abbreviated to **PDF** or pdf. The area under the graph of the PDF is also equal to 1.

A curved graph is sketched over each of the histograms in the diagram below.

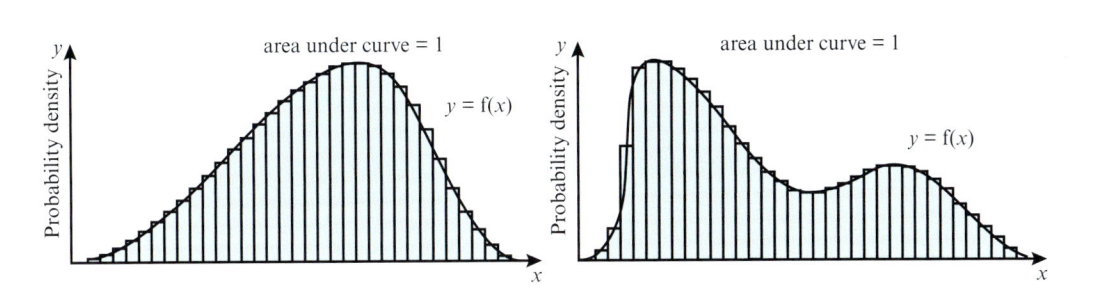

> **REWIND**
>
> We saw how to display continuous data in a histogram in Chapter 1, Section 1.3.

> **TIP**
>
> The word *function* should only be used when referring to a random variable. For data, we should rather use *curve* and/or *graph*.

If you were asked to describe these two curves, you may be tempted to say that the curve on the right is 'a bit odd' and that the curve on the left is 'a bit more normal'... and you would be quite right in doing so, as you will see shortly.

Three commonly occurring types of curved graph are shown in the diagrams below.

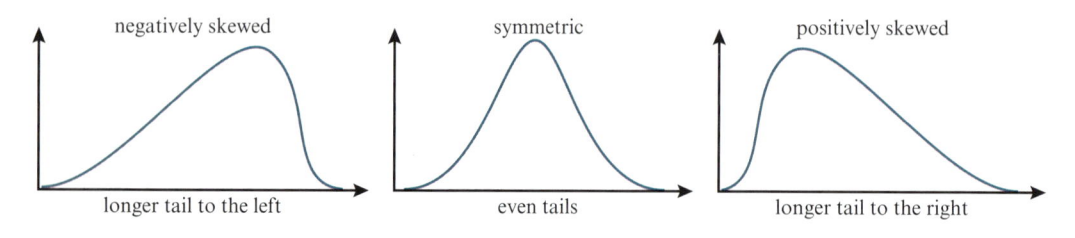

negatively skewed — longer tail to the left

symmetric — even tails

positively skewed — longer tail to the right

> **TIP**
>
> The mode is located at the graph's peak. The median is at the value where the area under the graph is divided into two equal parts; this value can be found by calculation from the histogram or estimated from a cumulative frequency graph.

EXPLORE 8.1

Three frequency distributions are shown in the tables below.

Use a histogram to sketch a graph representing the probability distribution for each of w, x and y.

w	$3 \leqslant w < 6$	$6 \leqslant w < 9$	$9 \leqslant w < 12$	$12 \leqslant w < 15$	$15 \leqslant w < 18$	$18 \leqslant w < 21$	$21 \leqslant w < 24$
f	13	13	13	13	13	13	13

x	$3 \leqslant x < 6$	$6 \leqslant x < 9$	$9 \leqslant x < 12$	$12 \leqslant x < 15$	$15 \leqslant x < 18$	$18 \leqslant x < 21$	$21 \leqslant x < 24$
f	3	9	18	24	18	9	3

y	$3 \leqslant y < 6$	$6 \leqslant y < 9$	$9 \leqslant y < 12$	$12 \leqslant y < 15$	$15 \leqslant y < 18$	$18 \leqslant y < 21$	$21 \leqslant y < 24$
f	8	19	10	4	10	19	8

Discuss and describe the shapes of the three graphs. What feature do they have in common?

Compare the measures of central tendency (averages) for w, x and y.

The normal curve

The frequency distribution of x in the Explore 8.1 activity produces a special type of curved graph. It is a symmetric, bell-shaped curve, known as a **normal curve**.

If a probability distribution is represented by a normal curve, then:

- Mean = median = mode
- The peak of the curve is at the mean (μ), and this is where we find the curve's line of symmetry
- Probability density decreases as we move away from the mean on both sides, so the further the values are from the mean, the less likely they are to occur
- An increase in the standard deviation (σ) means that values become more spread out from the mean. This results in the curve's width increasing and its height decreasing, so that the area under the graph is kept at a constant value of 1.

Graphs that represent probability distributions of related sets of data, such as the heights of the boys and the heights of the girls at your school, can be represented on the same diagram, so that comparisons can be made.

The following diagram shows two pairs of normal curves with their means and standard deviations compared. Note that the areas under the graphs in each pair are equal.

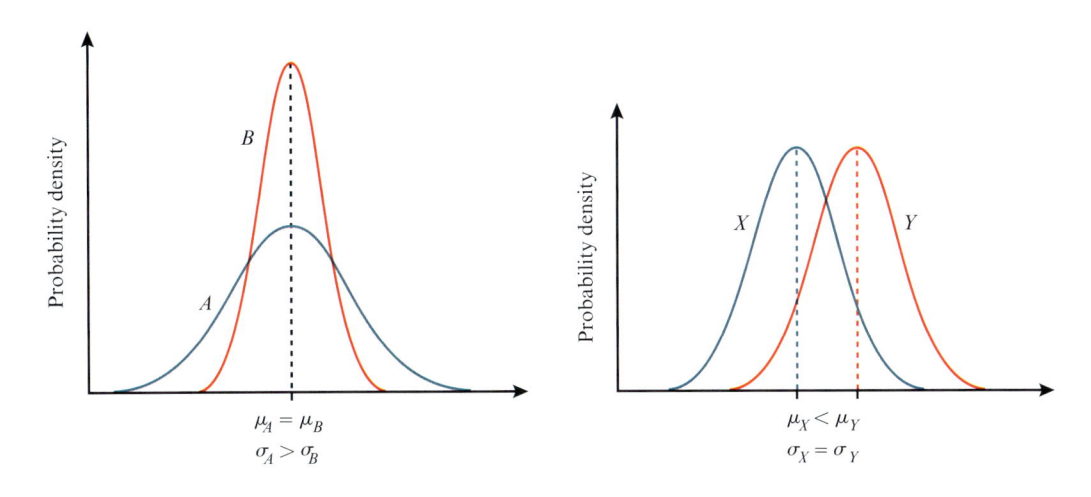

As we can see, A and B have the same mean, but the shapes of the normal curves are different because they do not have the same standard deviation. Curve B is obtained from curve A by stretching it both vertically (from the horizontal axis) and horizontally (from the line of symmetry).

X and Y have identically-shaped normal curves because they have the same standard deviation, but their positions or *locations* are different because they have different means. Each curve can be obtained from the other by a horizontal translation.

EXPLORE 8.2

You can investigate the effect of altering the mean and/or standard deviation on the location and shape of a normal curve by visiting the Density Curve of Normal Distribution resource on the GeoGebra website.

Note that the area under the curve is always equal to 1, whatever the values of μ and σ.

EXERCISE 8A

1 The probability distributions for A and B are represented in the diagram.

Indicate whether each of the following statements is true or false.

a $\mu_A > \mu_B$

b $\sigma_A < \sigma_B$

c A and B have the same range of values.

d $\sigma_A^2 = \sigma_B^2$

e At least half of the values in B are greater than μ_A.

f At most half of the values in A are less than μ_B.

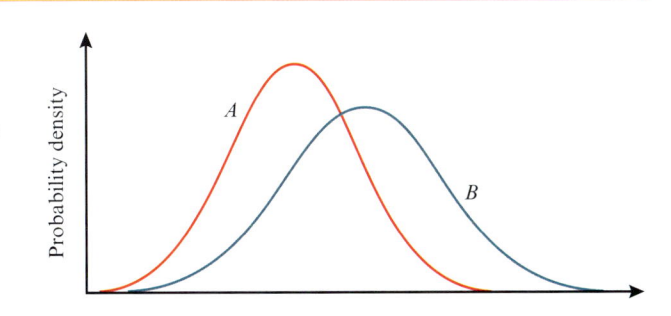

2 The diagram shows normal curves for the probability distributions of P and Q, that each contain n values.

   a Write down a statement comparing:

      i σ_P and σ_Q

      ii   the median value for P and the median value for Q

      iii the interquartile range for P and the interquartile range for Q.

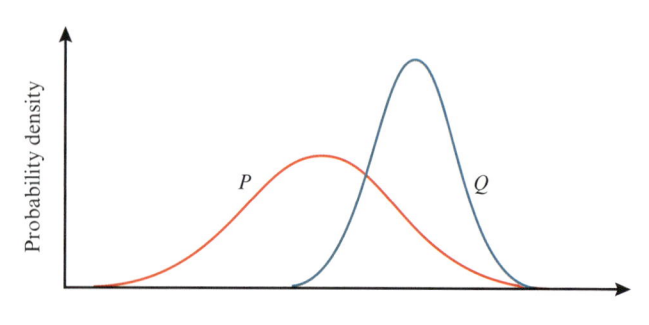

   b The datasets P and Q are merged to form a new dataset denoted by W.

      i   Describe the range of W.

      ii   Is the probability distribution for W a normal curve? Explain your answer.

      iii Copy the diagram above and sketch onto it a curved graph representing the probability distribution for W. Mark the relative positions of μ_P, μ_Q and μ_W along the horizontal axis.

3 The distributions of the heights of 1000 women and of 1000 men both produce normal curves, as shown. The mean height of the women is 160 cm and the mean height of the men is 180 cm.

   The heights of these women and men are now combined to form a new set of data. Assuming that the combined heights also produce a normal curve, copy the graph opposite and sketch onto it the curve for the combined heights of the 2000 women and men.

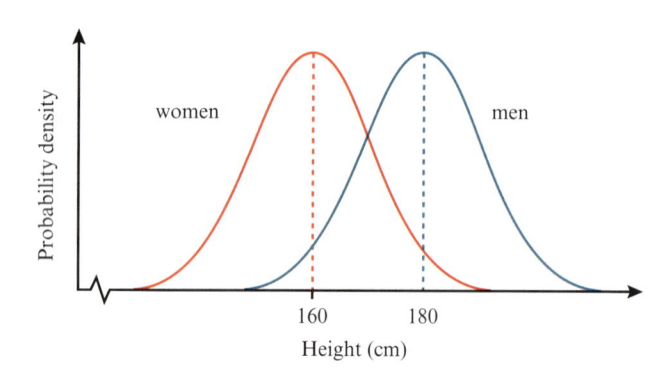

4 Probability distributions for the quantity of apple juice in 500 apple juice tins and for the quantity of peach juice in 500 peach juice tins are both represented by normal curves.

   The mean quantity of apple juice is 340 ml with variance 4 ml^2, and the mean quantity of peach juice is 340 ml with standard deviation 4 ml.

   a Copy the diagram and sketch onto it the normal curve for the quantity of peach juice in the peach juice tins.

   b Describe the curves' differences and similarities.

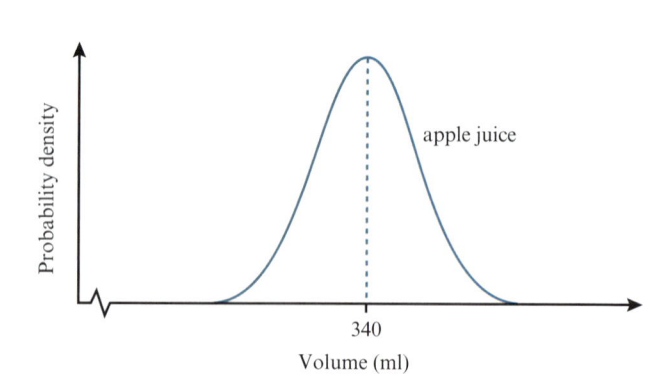

5 The masses of 444 newborn babies in the USA and 888 newborn babies in the UK both produce normal curves. For the USA babies, $\mu = 3.4$ kg and $\sigma = 200$ g; for the UK babies, $\mu = 3.3$ kg and $\sigma^2 = 36100$ g^2.

   a On a single diagram, sketch and label these two normal curves.

   b Describe the curves' differences and similarities.

6 The values in two datasets, whose probability distributions are both normal curves, are summarised by the following totals:

$\Sigma x^2 = 35\,000$, $\Sigma x = 12\,000$ and $n = 5000$.

$\Sigma y^2 = 72\,000$, $\Sigma y = 26\,000$ and $n = 10\,000$.

a Show that the centre of the curve for y is located to the right of the centre of the curve for x.

b On the same diagram, sketch a normal curve for each dataset.

8.2 The normal distribution

In Section 8.1, we saw how a curved graph can be used to represent the probability distribution of a set of continuous data. A curved graph that represents the probability distribution of a continuous random variable, as stated previously, is called a probability density function or PDF.

If we collect data on, say, the masses of a randomly selected sample of 1000 pineapples, we can produce a curved graph to illustrate the probabilities for the full and limited range of these masses. If there are no pineapples with masses under 0.2 kg or over 6 kg, then our graph will indicate that P(mass < 0.2) = 0 and P(mass > 6) = 0.

However, the continuous random variable 'the possible mass of a pineapple' is a theoretical model for the probability distribution. In the model, masses of less than 0.2 kg and masses of more than 6 kg would be shown to be extremely unlikely, but not impossible. The continuous random variable would, therefore, indicate that P(mass < 0.2) > 0 and P(mass > 6) > 0.

[Incidentally, the greatest ever recorded mass of a pineapple is 8.28 kg!]

The probability distribution of a continuous random variable is a mathematical function that provides a method of determining probabilities for the occurrence of different outcomes or observations.

E If the random variable X is normally distributed with mean μ and variance σ^2, then its equation is

$$f(x) = \frac{1}{\sigma\sqrt{2\pi}} \exp\left\{-\frac{(x-\mu)^2}{2\sigma^2}\right\}, \text{ for all real values of } x.$$

The parameters that define a normally distributed random variable are its mean μ and its variance σ^2.

To describe the normally distributed random variable X, we write $X \sim N(\mu, \sigma^2)$.

> **TIP**
>
> $\exp\{\ \}$ means the number $e = 2.71828\ldots$, raised to the power in the bracket, and $e^p > 0$ for any power p.

> **KEY POINT 8.1**
>
> $X \sim N(\mu, \sigma^2)$ describes a normally distributed random variable.
>
> We read this as 'X has a normal distribution with mean μ and variance σ^2'

The probability that X takes a value between a and b is equal to the area under the curve between the x-axis and the boundary lines $x = a$ and $x = b$.

The area under the graph of $y = f(x)$ can be found by integration: $P(a \leqslant X \leqslant b) = \int_a^b f(x)\,dx$

> **TIP**
>
> The area under any part of the curve is the same, whether or not the boundary values are included. $P(3 \leqslant X \leqslant 7)$, $P(3 \leqslant X < 7)$, $P(3 < X \leqslant 7)$ and $P(3 < X < 7)$ are indistinguishable.

Unfortunately, it is not possible to perform this integration accurately but, as we will see later, mathematicians have found ways to handle this challenge.

Normal distributions have many interesting properties, some of which are detailed in the following table.

Properties	Probabilities
Half of the values are less than the mean. Half of the values are greater than the mean.	$P(X < \mu) = P(X \leq \mu) = 0.5$ $P(X > \mu) = P(X \geq \mu) = 0.5$
Approximately 68.26% of the values lie within 1 standard deviation of the mean.	$P(\mu - \sigma \leq X \leq \mu + \sigma) = 0.6826$
Approximately 95.44% of the values lie within 2 standard deviations of the mean.	$P(\mu - 2\sigma \leq X \leq \mu + 2\sigma) = 0.9544$
Approximately 99.72% of the values lie within 3 standard deviations of the mean.	$P(\mu - 3\sigma \leq X \leq \mu + 3\sigma) = 0.9972$

> **TIP**
>
> The probability that the values in a normal distribution lie within a certain number of standard deviations of the mean is fixed.

In the following diagrams, the values 0, ±1 and ±2 represent numbers of standard deviations from the mean.

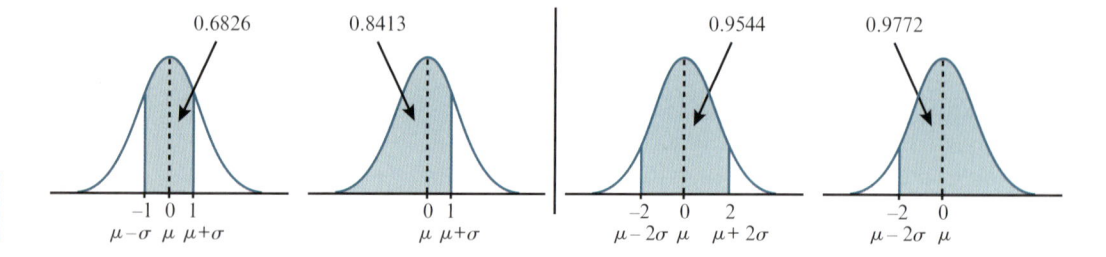

We can use the curve's symmetry, along with the table and diagrams above, to find other probabilities, such as:

We know that $P(-1 \leq X \leq 1) = 0.6826$, so

$P(X \leq 1) = \left(\frac{1}{2} \times 0.6826\right) + 0.5 = 0.8413 = P(X \geq -1)$.

We know that $P(-2 \leq X \leq 2) = 0.9544$, so

$P(X \leq 2) = \left(\frac{1}{2} \times 0.9544\right) + 0.5 = 0.9772 = P(X \geq -2)$.

EXPLORE 8.3

Calculated estimates of the mean and variance of the continuous random variables A, B and C are given in the following table.

	A	B	C
Mean	40	72	123
Variance	64	144	121

Random observations from each distribution were made with the following results:

For *A*: 8060 out of 13 120 observations lie in the interval from 32 to 48.

For *B*: 8475 out of 12 420 observations lie in the interval from 60 to 84.

For *C*: 8013 out of 10 974 observations lie in the interval from 112 to 134.

Investigate this information (using the previous table showing properties and probabilities for normal distributions) and comment on the statement 'The distributions of *A*, *B* and *C* are all normal'.

The standard normal variable Z

There are clearly an infinite number of values for the parameters of a normally distributed random variable. Nevertheless, most problems can be solved by transforming the random variable into a **standard normal variable**, which is denoted by Z, and which has mean 0 and variance 1.

By substituting $\mu = 0$ and $\sigma^2 = 1$ into the equation for the normal distribution PDF, we can find the equation of the PDF for $Z \sim N(0, 1)$. This is denoted by $\phi(z)$ and its equation is $\phi(z) = \frac{1}{\sqrt{2\pi}} \exp\left\{-\frac{z^2}{2}\right\}$. The graph of $y = \phi(z)$ is shown below.

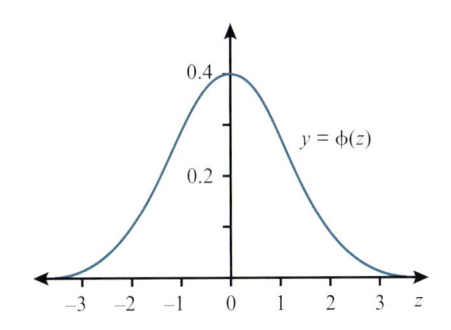

> **FAST FORWARD**
>
> Later in this section, we will see how any normal variable can be transformed to the standard normal variable by coding.

> **KEY POINT 8.2**
>
> The standard normal variable is $Z \sim N(0, 1)$

> **TIP**
>
> ϕ and Φ are the lower and upper-case Greek letter *phi*.

The mean of *Z* is 0.

The axis of symmetry is a vertical line through the mean, as with every normal distribution.

Z has a variance of 1 and, therefore, a standard deviation of 1.

$z = \pm 1, \pm 2$ and ± 3 represent values that are 1, 2 and 3 standard deviations above or below the mean.

Any $z < 0$ represents a value that is less the mean.

Any $z > 0$ represents a value that is greater the mean.

For $z > 3$ and for $z < -3$, $\phi(z) \approx 0$.

The area under the graph of $y = \phi(z)$ is equal to 1.

A vertical line drawn at any value of *Z* divides the area under the curve into two parts: one representing $P(Z \leq z)$ and the other representing $P(Z > z)$.

The value of $P(Z \leq z)$ is denoted by $\Phi(z)$ and, as mentioned earlier, we do not find such values by integration. Tables showing the value of $\Phi(z)$ for different values of *z* have been compiled and appear in the Standard normal distribution function table at the end of the book. In addition, some modern calculators are able to give the value of $\Phi(z)$ and the inverse function $\Phi^{-1}(z)$ directly.

Although only zero and positive values of Z (i.e. $z \geq 0$) appear in the tables, the graph's symmetry allows us to use the tables for positive and for negative values of z, as you will see after Worked example 8.2.

Values of the standard normal variable appear as 4-figure numbers from $z = 0.000$ to $z = 2.999$ in the tables. The first and second figures of z appear in the left-hand column; the third and fourth figures appear in the top row. The numbers in the 'ADD' column for the fourth figures indicate what we should add to the value of $\Phi(z)$ in the body of the table.

$\Phi(z)$ can be found for any given value of z, and z can be found for any given value of $\Phi(z)$ by using the tables in reverse (as shown in Worked example 8.4). In the critical values table, values for $\Phi(z)$ are denoted by p.

A section of the tables, from which we will find the value of $\Phi(0.274)$, is shown below.

> **TIP**
>
> *Critical values* refer to probabilities of 75%, 90%, 95%, ... and their complements 25%, 10%, 5%, ... and so on.

First and second figures | Third figure | Fourth figure

z	0	1	2	3	4	5	6	7	8	9	1	2	3	4	5	6	7	8	9
															ADD				
0.0	0.5000	0.5040	0.5080	0.5120	0.5160	0.5199	0.5239	0.5279	0.5319	0.5359	4	8	12	16	20	24	28	32	36
0.1	0.5398	0.5438	0.5478	0.5517	0.5557	0.5596	0.5636	0.5675	0.5714	0.5753	4	8	12	16	20	24	28	32	36
0.2	0.5793	0.5832	0.5871	0.5910	0.5949	0.5987	0.6026	0.6064	0.6103	0.6141	4	8	12	15	19	23	27	31	35
0.3	0.6179	0.6217	0.6255	0.6293	0.6331	0.6368	0.6406	0.6443	0.6480	0.6517	4	7	11	15	19	22	26	30	34

We locate the first and second figures of z (namely 0.2) in the left-hand column.

We then locate the third figure of z (namely 7) along the top row... this tells us that $\Phi(0.27) = 0.6064$.

Next we locate the fourth figure of z (namely 4) at the top-right. In line with 0.6064, we see 'ADD 15', which means that we must add 15 to the last two figures of 0.6064 to obtain the value of $\Phi(0.274)$.

$\Phi(0.274) = 0.6064 + 0.0015 = 0.6079$

> **TIP**
>
> $\Phi(0.274) = 0.6079$ can be expressed using inverse notation as $\Phi^{-1}(0.6079) = 0.274$.

WORKED EXAMPLE 8.1

Given that $Z \sim N(0, 1)$, find $P(Z < 1.23)$ and $P(Z \geq 1.23)$.

Answer

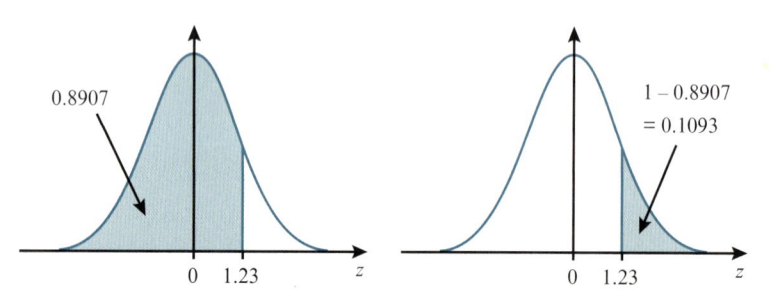

$\Phi(1.23) = 0.8907$ is the area to the left of $z = 1.23$.

$1 - \Phi(1.23) = 0.1093$ is the area to the right of $z = 1.23$, as shown in the graphs.

$\therefore P(Z < 1.23) = 0.8907$ and $P(Z \geq 1.23) = 0.1093$

Chapter 8: The normal distribution

WORKED EXAMPLE 8.2

Given that $Z \sim N(0, 1)$, find $P(0.4 \leq Z < 1.7)$ correct to 3 decimal places.

Answer

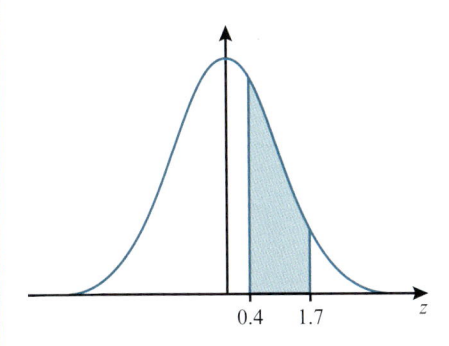

The required probability is equal to the difference between the area to the left of $z = 1.7$ and the area to the left of $z = 0.4$, as illustrated.

$\Phi(1.7) = 0.9554$ and $\Phi(0.4) = 0.6554$

We find the values of $\Phi(1.70)$ and $\Phi(0.40)$ in the main body of the table, which means that we do not need to use the ADD section.

$$P(0.4 \leq Z < 1.7) = P(Z < 1.7) - P(Z < 0.4)$$
$$= \Phi(1.7) - \Phi(0.4)$$
$$= 0.9554 - 0.6554$$
$$= 0.300$$

As noted previously, the normal distribution function tables do not show values for $z < 0$. However, we can use the symmetry properties of the normal curve, and the fact that the area under the curve is equal to 1, to find values of $\Phi(z)$ when z is negative.

Situations in which $z > 0$, and in which $z < 0$, are illustrated in the two diagrams below.

For a positive value, $z = b$:

The shaded area in this graph represents the value of $\Phi(b)$.

$\Phi(b) = P(Z \leq b)$

$\Phi(b) = P(Z \leq b)$ and $1 - \Phi(b) = P(Z \geq b)$.

For a negative value, $z = -a$:

The shaded area in this graph represents the value of $\Phi(a)$.

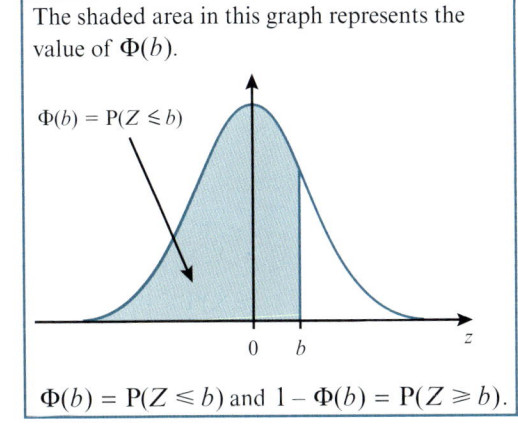

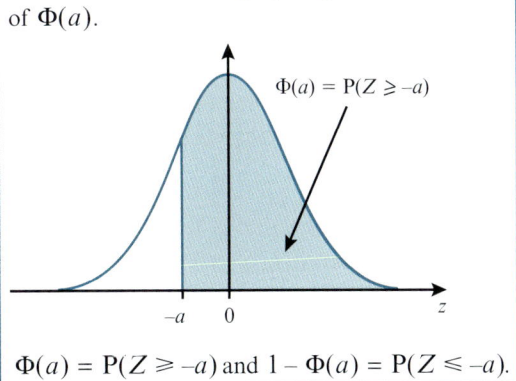

$\Phi(a) = P(Z \geq -a)$

$\Phi(a) = P(Z \geq -a)$ and $1 - \Phi(a) = P(Z \leq -a)$.

From the tables, the one piece of information, $\Phi(0.11) = 0.5438$, actually tells us four probabilities:

$P(Z \leq 0.11) = 0.5438$ and $P(Z \geq -0.11) = 0.5438$.

$P(Z \geq 0.11) = 1 - 0.5438 = 0.4562$ and $P(Z \leq -0.11) = 1 - 0.5438 = 0.4562$.

Information given about probabilities in a normal distribution should always be transferred to a sketched graph. Useful information, such as whether a particular value of z is positive or negative, will then be easy to see. This could, of course, also be determined by considering inequalities.

If, for example, $P(Z \geq z) > 0.5$, then $P(Z \leq z) < 0.5$ and, therefore, $z < 0$.

WORKED EXAMPLE 8.3

Given that $Z \sim N(0, 1)$, find $P(-1 \leq Z < 2.115)$ correct to 3 significant figures.

Answer

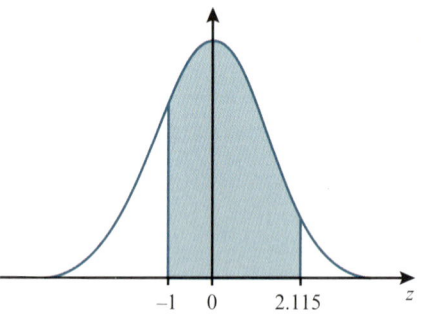

The required probability is given by the difference between the area to the left of $z = 2.115$ and the area to the left of $z = -1$.

The first of these areas is greater than 0.5 and the second is less than 0.5.

$P(Z < 2.115) = \Phi(2.115)$ and $P(Z < -1) = 1 - \Phi(1)$.

$$P(-1 \leq Z < 2.115) = \Phi(2.115) - [1 - \Phi(1)]$$
$$= \Phi(2.115) + \Phi(1) - 1$$
$$= 0.9828 + 0.8413 - 1$$
$$= 0.824$$

WORKED EXAMPLE 8.4

Given that $Z \sim N(0, 1)$, find the value of a such that $P(Z < a) = 0.9072$.

Answer

$0.9066 = \Phi(1.32)$

To find a value of z, we use the tables in reverse and search for the $\Phi(z)$ value closest to 0.9072, which is 0.9066.

$a = 1.32 + 0.004$
$ = 1.324$

For our value of 0.9072, we need to add 0.0006 to 0.9066, so 'ADD 6' is required – this will be done if 1.32 is given a 4th figure of 4.

$a = \Phi^{-1}(0.9072)$
$ = \Phi^{-1}(0.9066 + 0.0006)$
$ = \Phi^{-1}(0.9066) + 0.004$
$ = 1.32 + 0.004$
$ = 1.324$

We can check the value obtained for a by reading the tables in the usual way.
$\Phi(1.324) = \Phi(1.32) +$ 'ADD 6'
$ = 0.9066 + 0.0006$
$ = 0.9072$

Alternatively, we can show our working using inverse notation.

Chapter 8: The normal distribution

WORKED EXAMPLE 8.5

Given that $Z \sim N(0, 1)$, find the value of b such that $P(Z \geq b) = 0.7713$.

Answer

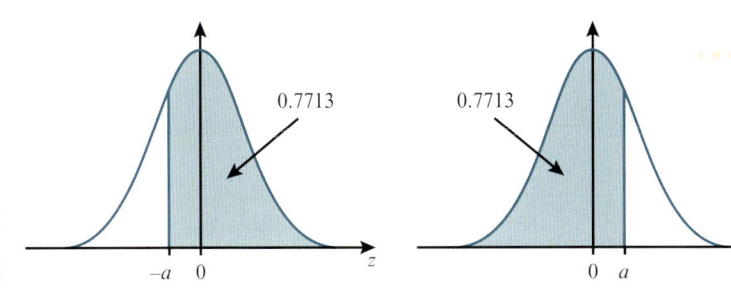

$P(Z \leq a) = 0.7713$, so $a = \Phi^{-1}(0.7713)$

$ = \Phi^{-1}(0.7704 + 0.0009)$

$ = \Phi^{-1}(0.7704) + 0.003$

$ = 0.740 + 0.003$

$ = 0.743$

$\therefore b = -a = -0.743$

$P(Z \geq b) > 0.5$ tells us that b is negative, so on our diagram we can replace b by $-a$, resulting in the two situations shown.

The value closest to 0.7713 in the tables is 0.7704. This requires the addition of 0.0009 to bring it up to 0.7713, and 9 is in the column headed 'ADD 3'.

EXERCISE 8B

1. Given that $Z \sim N(0, 1)$, find the following probabilities correct to 3 significant figures.
   a $P(Z < 0.567)$
   b $P(Z \leq 2.468)$
   c $P(Z > -1.53)$
   d $P(Z \geq -0.077)$
   e $P(Z > 0.817)$
   f $P(Z \geq 2.009)$
   g $P(Z < -1.75)$
   h $P(Z \leq -0.013)$
   i $P(Z < 1.96)$
   j $P(Z > 2.576)$

2. The random variable Z is normally distributed with mean 0 and variance 1. Find the following probabilities, correct to 3 significant figures.
   a $P(1.5 < Z < 2.5)$
   b $P(0.046 < Z < 1.272)$
   c $P(1.645 < Z < 2.326)$
   d $P(-2.807 < Z < -1.282)$
   e $P(-1.777 < Z < -0.746)$
   f $P(-1.008 < Z < -0.337)$
   g $P(-1.2 < Z < 1.2)$
   h $P(-1.667 < Z < 2.667)$
   i $P\left(-\frac{3}{4} \leq Z < \frac{8}{5}\right)$
   j $P\left(\sqrt{2} \leq Z < \sqrt{5}\right)$

3. Given that $Z \sim N(0, 1)$, find the value of k, given that:
   a $P(Z < k) = 0.9087$
   b $P(Z < k) = 0.5442$
   c $P(Z > k) = 0.2743$
   d $P(Z > k) = 0.0298$
   e $P(Z < k) = 0.25$
   f $P(Z < k) = 0.3552$
   g $P(Z > k) = 0.9296$
   h $P(Z > k) = 0.648$
   i $P(-k < Z < k) = 0.9128$
   j $P(-k < Z < k) = 0.6994$

4 Find the value of c in each of the following where Z has a normal distribution with $\mu = 0$ and $\sigma^2 = 1$.

a $P(c < Z < 1.638) = 0.2673$
b $P(c < Z < 2.878) = 0.4968$
c $P(1 < Z < c) = 0.1408$
d $P(0.109 < Z < c) = 0.35$
e $P(c < Z < 2) = 0.6687$
f $P(c < Z < 1.85) = 0.9516$
g $P(-1.221 < Z < c) = 0.888$
h $P(-0.674 < Z < c) = 0.725$
i $P(-2.63 < Z < c) = 0.6861$
j $P(-2.7 < Z < c) = 0.0252$

Standardising a normal distribution

The probability distribution of a normally distributed random variable is represented by a normal curve. This curve is centred on the mean μ; the area under the curve is equal to 1, and its height is determined by the standard deviation σ.

We already have a method for finding probabilities involving the standard normal variable $Z \sim N(0, 1)$ using the normal distribution function tables. Fortunately, this same set of tables can be used to find probabilities involving any normal random variable, no matter what the values of μ and σ^2. Although we have only learnt about coding data, it turns out that coding works in exactly the same way for normally distributed random variables: they behave in the way that we expect and remain normal after coding.

If we code X by subtracting μ, then the PDF is translated horizontally by $-\mu$ units and is now centred on 0. The new random variable $X - \mu$ has mean 0 and standard deviation σ.

If we now code $X - \mu$ by multiplying by $\frac{1}{\sigma}$ (i.e. dividing by σ) then the standard deviation (and variance) will be equal to 1, while the mean remains 0.

The coded random variable $\frac{X - \mu}{\sigma}$ is normally distributed with mean 0 and variance 1.

Coding the random variable X in this way is called *standardising*, because it transforms the distribution $X \sim N(\mu, \sigma^2)$ to $Z \sim N(0, 1)$.

> **REWIND**
>
> In Chapter 2, Section 2.2 and in Chapter 3, Section 3.3, we saw how the coding of data by addition and/or multiplication affects the mean and the standard deviation.

> **FAST FORWARD**
>
> We will learn more about coding random variables in the Probability & Statistics 2 Coursebook, Chapter 3.

> **REWIND**
>
> In the table showing properties and probabilities of normal distributions prior to Explore 8.3, we saw that probabilities are determined by the number of standard deviations from the mean. The properties given in that table apply to all normal random variables.

KEY POINT 8.3

When $X \sim N(\mu, \sigma^2)$ then $Z = \frac{X - \mu}{\sigma}$ has a standard normal distribution.

A standardised value $z = \frac{x - \mu}{\sigma}$ tells us how many standard deviations x is from the mean.

Probabilities involving values of X are equal to probabilities involving the corresponding values of Z, which can be found from the normal distribution function tables for $Z \sim N(0, 1)$.

For example, if $X \sim N(20, 9)$, then $P(X < 23) = P\left(Z < \frac{23 - 20}{\sqrt{9}}\right)$.

Chapter 8: The normal distribution

WORKED EXAMPLE 8.6

Given that $X \sim N(11, 25)$, find $P(X < 18)$ correct to 3 significant figures.

Answer

$$z = \frac{18-11}{\sqrt{25}} = 1.4$$

We standardise $x = 18$ and find that it is 1.4σ above the mean of 11.

$$P(X < 18) = P(Z < 1.4)$$
$$= \Phi(1.4)$$
$$= 0.919$$

WORKED EXAMPLE 8.7

Given that $X \sim N(20, 7)$, find $P(X \leq 16.6)$ correct to 3 significant figures.

Answer

$$z = \frac{16.6 - 20}{\sqrt{7}} = -1.285$$

We standardise $x = 16.6$ and find that it is 1.285σ below the mean of 20.

$$P(X \leq 16.6) = P(Z \leq -1.285)$$
$$= 1 - \Phi(1.285)$$
$$= 0.0994$$

WORKED EXAMPLE 8.8

Given that $X \sim N(5, 5)$, find $P(2 \leq X < 9)$ correct to 3 significant figures.

Answer

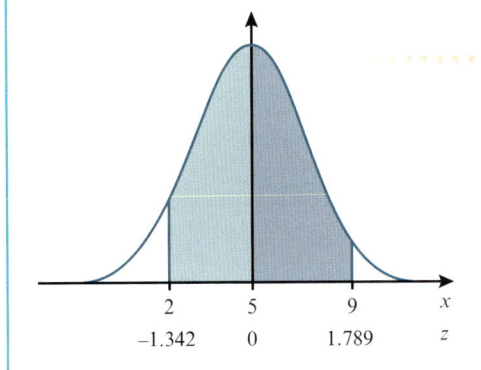

For $x = 2$, $z = \dfrac{2-5}{\sqrt{5}} = -1.342$

For $x = 9$, $z = \dfrac{9-5}{\sqrt{5}} = 1.789$

The required area is shown in two parts in the diagram.

Area to the right of $z = 0$ is
$\Phi(1.789) - \Phi(0) = \Phi(1.789) - 0.5$
$= 0.4633$

Area to the left of $z = 0$ is
$\Phi(0) - \Phi(-1.342) = 0.5 - [1 - \Phi(1.342)]$
$= 0.4102$
Total area $= 0.4633 + 0.4102 = 0.8735$
$\therefore P(2 \leq X < 9) = 0.874$

Here, we find the two areas separately then add them to obtain our final answer, which is where we round to the accuracy specified in the question.

Try solving this problem by the method shown in Worked example 8.3 (using the areas to the left of both $z = 1.789$ and $z = -1.342$) and decide which approach you prefer.

TIP

Where possible, always use a 4-figure value for z.

Some useful results from previous worked examples are detailed in the following graphs.

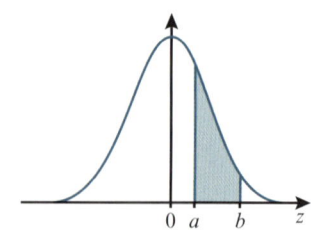

For $0 < a < b$

$P(a < Z < b) = \Phi(b) - \Phi(a)$

For $-a < 0 < b$

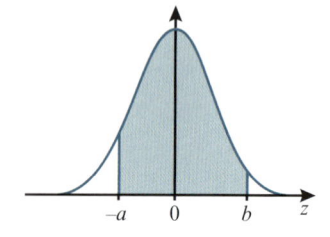

$P(-a < Z < b) = \Phi(b) + \Phi(a) - 1$

For $-a < 0 < a$

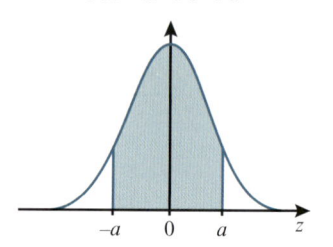

$P(-a < Z < a) = 2\Phi(a) - 1$

WORKED EXAMPLE 8.9

Given that $Y \sim N(\mu, \sigma^2)$, $P(Y < 10) = 0.75$ and $P(Y \geq 12) = 0.1$, find the values of μ and σ.

Answer

$P(Y \geq 12) < 0.5$, so $P(Y < 12) > 0.5$, which means that $12 > \mu$.

$P(Y < 10) > 0.5$, which means also that $10 > \mu$.

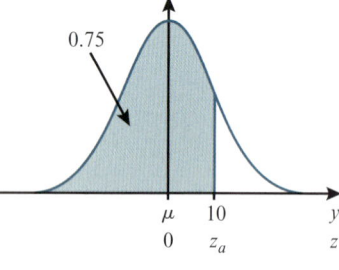

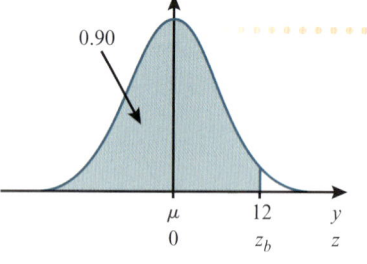

These simple sketch graphs allow us to locate the values 10 and 12 relative to μ.

$\dfrac{10 - \mu}{\sigma} = 0.674$ gives $10 - \mu = 0.674\sigma$ ...... [1]

$\dfrac{12 - \mu}{\sigma} = 1.282$ gives $12 - \mu = 1.282\sigma$ ...... [2]

$z_a = \Phi^{-1}(0.75) = 0.674$ and
$z_b = \Phi^{-1}(0.90) = 1.282$

Note that both 0.75 and 0.90 are critical values.

Chapter 8: The normal distribution

$12 - \mu = 1.282\sigma$ [2]

$10 - \mu = 0.674\sigma$ [1]

$\overline{ 2 } = 0.608\sigma$

$\therefore \sigma = 3.29$ and $\mu = 7.78$.

> We subtract equation [1] from [2] to solve this pair of simultaneous equations.

EXERCISE 8C

1 Standardise the appropriate value(s) of the normal variable X represented in each diagram, and find the required probabilities correct to 3 significant figures.

  a Find $P(X \leq 11)$, given that $X \sim N(8, 25)$.

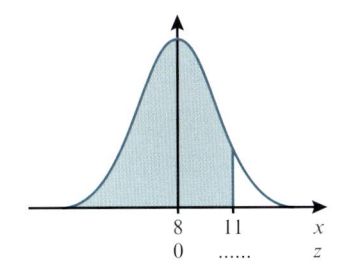

  b Find $P(X < 69.1)$, given that $X \sim N(72, 11)$.

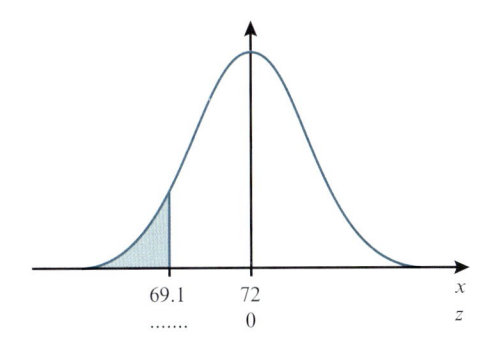

  c Find $P(3 < X < 7)$, given that $X \sim N(5, 5)$.

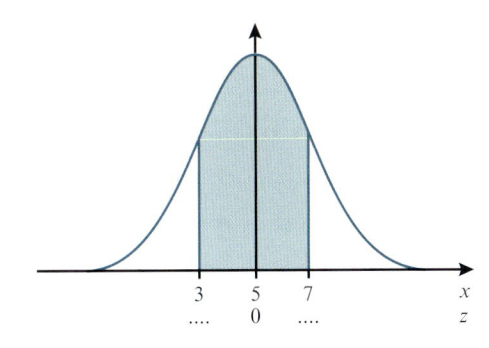

From this point in the exercise, you are strongly advised to sketch a diagram to help answer each question.

2 Calculate the required probabilities correct to 3 significant figures.
   a Find $P(X \leq 9.7)$ and $P(X > 9.7)$, given that $X \sim N(6.2, 6.25)$.
   b Find $P(X \leq 5)$ and $P(X > 5)$, given that $X \sim N(3, 49)$.
   c Find $P(X > 33.4)$ and $P(X \leq 33.4)$, given that $X \sim N(37, 4)$.
   d Find $P(X < 13.5)$ and $P(X \geq 13.5)$, given that $X \sim N(20, 15)$.
   e Find $P(X > 91)$ and $P(X \leq 91)$, given that $X \sim N(80, 375)$.
   f Find $P(1 \leq X < 21)$, given that $X \sim N(11, 25)$.
   g Find $P(2 \leq X < 5)$, given that $X \sim N(3, 7)$.
   h Find $P(6.2 \geq X \geq 8.8)$, given that $X \sim N(7, 1.44)$. [Read carefully.]
   i Find $P(26 \leq X < 28)$, given that $X \sim N(25, 6)$.
   j Find $P(8 \leq X < 10)$, given that $X \sim N(12, 2.56)$.

3 a Find a, given that $X \sim N(30, 16)$ and that $P(X \leq a) = 0.8944$.
   b Find b, given that $X \sim N(12, 4)$ and that $P(X \leq b) = 0.9599$.
   c Find c, given that $X \sim N(23, 9)$ and that $P(X > c) = 0.9332$.
   d Find d, given that $X \sim N(17, 25)$ and that $P(X > d) = 0.0951$.
   e Find e, given that $X \sim N(100, 64)$ and that $P(X > e) = 0.95$.

4 a Find f, given that $X \sim N(10, 7)$ and that $P(f \leq X < 13.3) = 0.1922$.
   b Find g, given that $X \sim N(45, 50)$ and that $P(g \leq X < 55) = 0.5486$.
   c Find h, given that $X \sim N(7, 2)$ and that $P(8 \leq X < h) = 0.216$.
   d Find j, given that $X \sim N(20, 11)$ and that $P(j \leq X < 22) = 0.5$.

5 X is normally distributed with mean 4 and variance 6. Find the probability that X takes a negative value.

6 Given that $X \sim N\left(\mu, \frac{4}{9}\mu^2\right)$ where $\mu > 0$, find $P(X < 2\mu)$.

7 If $T \sim N(10, \sigma^2)$ and $P(T > 14.7) = 0.04$, find the value of σ.

8 It is given that $V \sim N(\mu, 13)$ and $P(V < 15) = 0.75$. Find the value of μ.

9 The variable $W \sim N(\mu, \sigma^2)$. Given that $\mu = 4\sigma$ and $P(W < 83) = 0.95$, find the value of μ and of σ.

10 X has a normal distribution in which $\sigma = \mu - 30$ and $P(X \geq 12) = 0.9$. Find the value of μ and of σ.

11 The variable $Q \sim N(\mu, \sigma^2)$. Given that $P(Q < 1.288) = 0.281$ and $P(Q < 6.472) = 0.591$, find the value of μ and of σ, and calculate $P(4 \leq Q < 5)$.

12 For the variable $V \sim N(\mu, \sigma^2)$, it is given that $P(V < 8.4) = 0.7509$ and $P(V > 9.2) = 0.1385$.
   Find the value of μ and of σ, and calculate $P(V \leq 10)$.

13 Find the value of μ and of σ and calculate $P(W > 6.48)$ for the variable $W \sim N(\mu, \sigma^2)$, given that $P(W \geq 4.75) = 0.6858$ and $P(W \leq 2.25) = 0.0489$.

14 X has a normal distribution, such that $P(X > 147.0) = 0.0136$ and $P(X \leq 59.0) = 0.0038$.
   Use this information to calculate the probability that $80.0 \leq X < 130.0$.

8.3 Modelling with the normal distribution

The German mathematician Carl Friedrich Gauss showed that measurement errors made in astronomical observations were well modelled by a normal distribution, and the Belgian statistician and sociologist Adolphe Quételet later applied this to human characteristics when he saw that distributions of such things as height, weight, girth and strength were approximately normal.

We are now in a position to apply our knowledge to real-life situations, and to solve more advanced problems involving the normal distribution.

WORKED EXAMPLE 8.10

The mass of a newborn baby in a certain region is normally distributed with mean 3.35 kg and variance 0.0858 kg². Estimate how many of the 1356 babies born last year had masses of less than 3.5 kg.

Answer

$\Phi(Z) = \Phi\left(\dfrac{3.5 - 3.35}{\sqrt{0.0858}}\right)$ ⋯⋯ We standardise the mass of 3.5 kg.

$= \Phi(0.512)$

$= 0.6957$ ⋯⋯ 0.6957 is a relative frequency equal to 69.57%.

P(mass < 3.5 kg) = 0.6957
69.57% of 1356 = 943.3692
∴ There were about 943 newborn babies.

> **TIP**
> We cannot know the exact number of newborn babies from the model because it only gives estimates. However, we do know that the number of babies must be an integer.

WORKED EXAMPLE 8.11

A factory produces half-litre tins of oil. The volume of oil in a tin is normally distributed with mean 506.18 ml and standard deviation 2.96 ml.

  a What percentage of the tins contain less than half a litre of oil?

  b Find the probability that exactly 1 out of 3 randomly selected tins contains less than half a litre of oil.

Answer

a $z = \dfrac{500 - 506.18}{2.96} = -2.088$ ⋯⋯ Let X represent the amount of oil in a tin, then $X \sim N(506.18, 2.96^2)$.

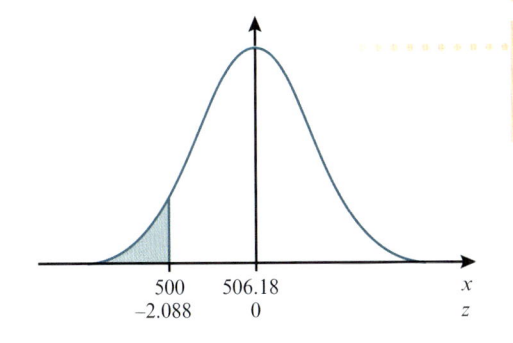

⋯⋯ The graph shows the probability distribution for the amount of oil in a tin.

$$P(X < 500) = P(Z < -2.088)$$
$$= 1 - \Phi(2.088)$$
$$= 1 - 0.9816$$
$$= 0.0184$$

∴ 1.84% of the tins contain less than half a litre of oil.

b $P(Y = 1) = \binom{3}{1} \times 0.0184^1 \times 0.9816^2$

$\qquad\qquad = 0.0532$

Let the discrete random variable Y be the number of tins containing less than half a litre of oil, then $Y \sim B(3, 0.0184)$.

> **TIP**
> A probability obtained from a normal distribution can be used as the parameter p in a binomial distribution.

EXERCISE 8D

1 The length of a bolt produced by a machine is normally distributed with mean 18.5 cm and variance 0.7 cm². Find the probability that a randomly selected bolt is less than 18.85 cm long.

2 The waiting times, in minutes, for patients at a clinic are normally distributed with mean 13 and variance 16.

  a Calculate the probability that a randomly selected patient has to wait for more than 16.5 minutes.

  b Last month 468 patients attended the clinic. Calculate an estimate of the number who waited for less than 9 minutes.

3 Tomatoes from a certain producer have masses which are normally distributed with mean 90 grams and standard deviation 17.7 grams. The tomatoes are sorted into three categories by mass, as follows:

  Small: under 80 g;   Medium: 80 g to 104 g;   Large: over 104 g.

  a Find, correct to 2 decimal places, the percentage of tomatoes in each of the three categories.

  b Find the value of k such that $P(k \leq X < 104) = 0.75$, where X is the mass of a tomato in grams.

4 The heights, in metres, of the trees in a forest are normally distributed with mean μ and standard deviation 3.6. Given that 75% of the trees are less than 10 m high, find the value of μ.

5 The mass of a certain species of fish caught at sea is normally distributed with mean 5.73 kg and variance 2.56 kg². Find the probability that a randomly selected fish caught at sea has a mass that is:

  a less than 6.0 kg   **b** more than 3.9 kg   **c** between 7.0 and 8.0 kg

6 The distance that children at a large school can hop in 15 minutes is normally distributed with mean 199 m and variance 3700 m².

  a Calculate an estimate of b, given that only 25% of the children hopped further than b metres.

  b Find an estimate of the interquartile range of the distances hopped.

7 The daily percentage change in the value of a company's shares is expected to be normally distributed with mean 0 and standard deviation 0.51. On how many of the next 365 working days should the company expect the value of its shares to fall by more than 1%?

8 The masses, w grams, of a large sample of apples are normally distributed with mean 200 and variance 169. Given that the masses of 3413 apples are in the range $187 \leq w < 213$, calculate an estimate of the number of apples in the sample.

9 The ages of the children in a gymnastics club are normally distributed with mean 15.2 years and standard deviation σ. Find the value of σ given that 30.5% of the children are less than 13.5 years of age.

10 The speeds, in kmh^{-1}, of vehicles passing a particular point on a rural road are normally distributed with mean μ and standard deviation 20. Find the value of μ and find what percentage of the vehicles are being driven at under 80 kmh^{-1}, given that 33% of the vehicles are being driven at over 100 kmh^{-1}.

11 Coffee beans are packed into bags by the workers on a farm, and each bag claims to contain 200 g. The actual mass of coffee beans in a bag is normally distributed with mean 210 g and standard deviation σ. The farm owner informs the workers that they must repack any bag containing less than 200 g of coffee beans. Find the value of σ, given that 0.5% of the bags must be repacked.

12 Colleen exercises at home every day. The length of time she does this is normally distributed with mean 12.8 minutes and standard deviation σ. She exercises for more than 15 minutes on 42 days in a year of 365 days.

   a Calculate the value of σ.

   b On how many days in a year would you expect Colleen to exercise for less than 10 minutes?

13 The times taken by 15-year-olds to solve a certain puzzle are normally distributed with mean μ and standard deviation 7.42 minutes.

   a Find the value of μ, given that three-quarters of all 15-year-olds take over 20 minutes to solve the puzzle.

   b Calculate an estimate of the value of n, given that 250 children in a random sample of n 15-year-olds fail to solve the puzzle in less than 30 minutes.

14 The lengths, X cm, of the leaves of a particular species of tree are normally distributed with mean μ and variance σ^2.

   a Find $P(\mu - \sigma \leq X < \mu + \sigma)$.

   b Find the probability that a randomly selected leaf from this species has a length that is more than 2 standard deviations from the mean.

   c Find the value of μ and of σ, given that $P(X < 7.5) = 0.75$ and $P(X < 8.5) = 0.90$.

15 The time taken in seconds for Ginger's computer to open a specific large document is normally distributed with mean 9 and variance 5.91.

   a Find the probability that it takes exactly 5 seconds or more to open the document.

   b Ginger opens the document on her computer on n occasions. The probability that it fails to open in less than exactly 5 seconds on at least one occasion is greater than 0.5. Find the least possible value of n.

16 The masses of all the different pies sold at a market are normally distributed with mean 400 g and standard deviation 61 g. Find the probability that:

   a the mass of a randomly selected pie is less than 425 g

   b 4 randomly selected pies all have masses of less than 425 g

   c exactly 7 out of 10 randomly selected pies have masses of less than 425 g.

17 The height of a female university student is normally distributed with mean 1.74 m and standard deviation 12.3 cm. Find the probability that:

   a a randomly selected female student is between 1.71 and 1.80 metres tall

   b 3 randomly selected female students are all between 1.71 and 1.80 m tall

   c exactly 15 out of 50 randomly selected female students are between 1.71 and 1.80 metres tall.

8.4 The normal approximation to the binomial distribution

In Chapter 7, Section 7.1, we saw that the binomial distribution can be used to solve problems such as 'Find the probability of obtaining exactly 60 heads with 100 tosses of a fair coin', and that this is equal to $\binom{100}{60} \times 0.5^{60} \times 0.5^{40}$. Therefore, to find the probability of obtaining 60 or more heads, we must find the probability for 60 heads, for 61 heads, for 62 heads and so on, and add them all together.

Imagine how long it took to calculate binomial probabilities before calculators and computers!

However, in certain situations, we can approximate a probability such as this by a method that involves far fewer calculations using the normal distribution.

EXPLORE 8.4

Binomial probability distributions for 2, 4, and 12 tosses of a fair coin are shown in the following diagrams. Notice that, as the number of coin tosses increases, the shape of the probability distribution becomes increasingly normal.

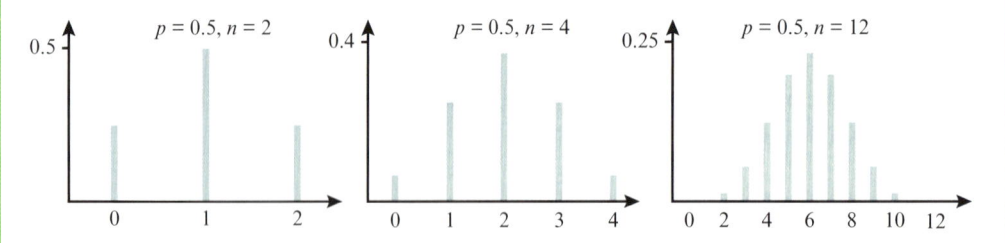

Does the binomial probability distribution maintain its normal shape for large values of n when p varies? Find out using the Binomial Distribution resource on the GeoGebra website.

Select any $n \geqslant 20$ then use the pause/play button or the slider to vary the value of p. Take note of when the distribution loses its normal shape. Repeat this for other values of n.

Can you generalise as to when the binomial distribution begins to lose its normal shape?

DID YOU KNOW?

Abraham de Moivre, the 18th century statistician and consultant, was often asked to make long calculations concerning games of chance. He noted that when the number of events increased, the shape of the binomial distribution approached a very smooth curve, and saw that he would be able to solve these long calculation problems if he could find a mathematical expression for this curve: this is exactly what he did. The curve he discovered is now called the normal curve.

Before the late 1870s, when the term *normal* was coined independently by Peirce, Galton and Lexis, this distribution was known – and still is by some – as the Gaussian distribution after the German mathematician Carl Friedrich Gauss. The word *normal* is not meant to suggest that all other distributions are abnormal!

Abraham de Moivre 1667–1754

FAST FORWARD

de Moivre's theorem, $(\cos\theta + i\sin\theta)^n = \cos n\theta + i\sin n\theta$ links trigonometry with complex numbers – a topic that we cover in the Pure Mathematics 2 & 3 Coursebook, Chapter 11.

The following diagrams show the shapes of four binomial distributions for $n = 25$.

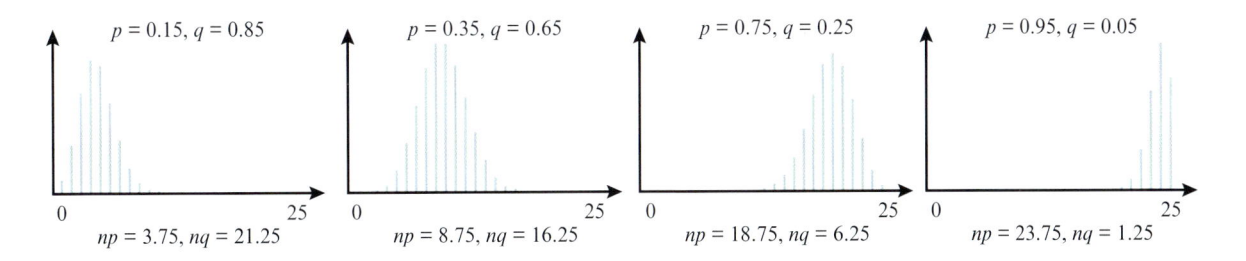

$p = 0.15, q = 0.85$	$p = 0.35, q = 0.65$	$p = 0.75, q = 0.25$	$p = 0.95, q = 0.05$
$np = 3.75, nq = 21.25$	$np = 8.75, nq = 16.25$	$np = 18.75, nq = 6.25$	$np = 23.75, nq = 1.25$

As you can see, the binomial distribution loses its normal shape when p is small and also when q is small.

A more detailed investigation shows that the binomial distribution has an approximately normal shape if np and nq are both greater than 5. These are the values that we use to decide whether a binomial distribution can be well-approximated by a normal distribution. The larger the values of np and nq, the more accurate the approximation will be. As we can see from the above diagrams, the approximation is adequate (but not very good) when $np = 18.75$ and $nq = 6.25$.

The distribution $X \sim B(40, 0.9)$ cannot be well-approximated by a normal distribution because $nq < 5$.

The distribution $X \sim B(250, 0.2)$ can be well-approximated by a normal distribution because $np = 50$ and $nq = 200$, both of which are substantially greater than 5.

When we approximate a discrete distribution by a continuous distribution, a discrete value such as $X = 13$ must be treated as being represented by the class of continuous values $12.5 \leq X < 13.5$. For this reason, $X = 13$ must be replaced by either $X = 12.5$ or by $X = 13.5$ in our probability calculations. Making this replacement is known as 'making a continuity correction'. Deciding whether to use $X = 12.5$ or $X = 13.5$ depends on whether or not $X = 13$ is included in the probability that we wish to find.

For example, if we wish to find $P(X < 13)$, where $X = 13$ is not included, we calculate using $X = 12.5$.

If we wish to find $P(X \leq 13)$, where $X = 13$ is included, we calculate using $X = 13.5$.

Further details of continuity corrections are given in Worked example 8.12.

> **KEY POINT 8.4**
>
> $X \sim B(n, p)$ can be approximated by $N(\mu, \sigma^2)$, where $\mu = np$ and $\sigma^2 = npq$, provided that n is large enough to ensure that $np > 5$ and $nq > 5$.

> **KEY POINT 8.5**
>
> Continuity corrections must be made when a discrete distribution is approximated by a continuous distribution.

WORKED EXAMPLE 8.12

Given that $X \sim B(100, 0.4)$, use a suitable approximation and continuity correction to find:

a $P(X < 43)$

b $P(X > 43)$

Answer

$\mu = np = 40$ and $\sigma^2 = npq = 24$.
$X \sim B(100, 0.4)$ can be approximated by $N(40, 24)$.

> The conditions for approximating a binomial distribution by a normal distribution are met because $np = 40$ and $nq = 60$, which are both greater than 5.

> In the continuous distribution $N(40, 24)$, 43 is represented by the class of continuous values $42.5 \leq X < 43.5$, as shown in the diagram.

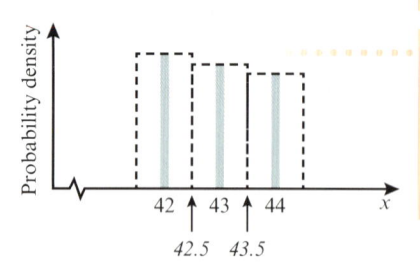

Possible continuity corrections for a discrete value of 43 are given below:

For $P(X < 43)$, we would use the lower boundary value 42.5 ...... [part a]
For $P(X \leq 43)$, we would use the upper boundary value 43.5

For $P(X > 43)$, we would use the upper boundary value 43.5 ...... [part b]
For $P(X \geq 43)$, we would use the lower boundary value 42.5

a $P(X < 43) \approx P(Z < 0.510)$
$= \Phi(0.510)$
$= 0.6950$
$\therefore P(X < 43) \approx 0.695$

> For $x = 42.5$, $z = \dfrac{42.5 - 40}{\sqrt{24}} = 0.5103$

b $P(X > 43) \approx P(Z > 0.714)$
$= 1 - \Phi(0.714)$
$= 0.2377$
$\therefore P(X > 43) \approx 0.238$

> For $x = 43.5$, $z = \dfrac{43.5 - 40}{\sqrt{24}} = 0.7144$

> **FAST FORWARD**
>
> We will also make continuity corrections when using the normal distribution as an approximation to the Poisson distribution in the Probability & Statistics 2 Coursebook, Chapter 2.

> **TIP**
>
> $X < a$ means 'X is fewer/less than a'.
>
> $X > a$ means 'X is more/greater than a'.
>
> $X \leq a$ means 'X is at most a' and 'X is not more than a' and 'X is a or less'.
>
> $X \geq a$ means 'X is at least a' and 'X is not less than a' and 'X is a or more'.

WORKED EXAMPLE 8.13

Boxes are packed with 8000 randomly selected items. It is known that 0.2% of the items are yellow.

Find, using a suitable approximation, the probability that:

a a box contains fewer than 20 yellow items

b exactly 2 out of 3 randomly selected boxes contain fewer than 20 yellow items.

Answer

a $X \sim B(8000, 0.002)$ Let X be the number of yellow items in a box.

$X \sim B(8000, 0.002)$ can be approximated by $N(16, 15.968)$.

$np = 16$ and $nq = 7984$ are both greater than 5, so we can approximate the binomial distribution by a normal distribution using $\mu = np = 16$ and $\sigma^2 = npq = 15.968$.

$$P(X < 20) \approx P\left(Z < \frac{19.5 - 16}{\sqrt{15.968}}\right)$$
$$= P(Z < 0.87587\ldots)$$
$$= \Phi(0.876)$$
$$= 0.8094$$

To find $P(X < 20)$, we must calculate with the value $x = 19.5$.

TIP

Do not forget to make the continuity correction!

∴ The probability that a box contains fewer than 20 yellow items is approximately 0.809.

b $P(Y = 2) = \binom{3}{2} \times 0.8094^2 \times 0.1906^1$
$= 0.375$

Let Y be the number of boxes containing fewer than 20 yellow items, then $Y \sim B(3, 0.8094)$.

TIP

Although our answer to part **a** is only an approximation, we should not use a rounded probability, such as 0.8, in further calculations.

WORKED EXAMPLE 8.14

A fair coin is tossed 888 times. Find, by use of a suitable approximation, the probability that the coin lands heads-up at most 450 times.

Answer

$X \sim B(888, 0.5)$ Let X represent the number of times the coin lands heads-up.

$X \sim B(888, 0.5)$ can be approximated by $N(444, 222)$.

$np = 444$, $nq = 444$ are both greater than 5, and $npq = 222$.

$$P(X \leq 450) \approx P\left(Z \leq \frac{450.5 - 444}{\sqrt{222}}\right)$$
$$= P(Z \leq 0.436)$$
$$= \Phi(0.436)$$
$$= 0.6686$$

To find $P(X \leq 450)$ we calculate with $x = 450.5$.

∴ P(at most 450 heads) ≈ 0.669

EXPLORE 8.5

By visiting the Binomial and Normal resource on the GeoGebra website you will get a clear picture of how the normal approximation to the binomial distribution works.

Select values of n and p so that np and $n(1-p)$ are both greater than 5.

The binomial probability distribution is displayed with an overlaid normal curve (the value of $\mu = np$ and of $\sigma = \sqrt{np(1-p)}$ are displayed in red at the top-right). If you then check the *probability box*, adjustable values of $x = a$ and $x = b$ appear on the diagram, with the area between them shaded. Remember that a discrete variable is being approximated by a continuous variable, so appropriate continuity corrections are needed to find the best probability estimates.

EXERCISE 8E

1. Decide whether or not each of the following binomial distributions can be well-approximated by a normal distribution.

    For those that can, state the values of the parameters μ and σ^2.

    For those that cannot, state the reason.

    a B(20, 0.6)   b B(30, 0.95)   c B(40, 0.13)   d B(50, 0.06)

2. Find the smallest possible value of n for which the following binomial distributions can be well-approximated by a normal distribution.

    a B(n, 0.024)   b B(n, 0.15)   c B(n, 0.52)   d B(n, 0.7)

3. Describe the binomial distribution that can be approximated by the normal distribution N(14, 10.5).

4. By first evaluating np and npq, use a suitable approximation and continuity correction to find $P(X < 75)$ for the discrete random variable $X \sim B(100, 0.7)$.

5. The discrete random variable $Y \sim B(50, 0.6)$. Use a suitable approximation and continuity correction to find $P(Y > 26)$.

6. A biased coin is tossed 160 times. The number of heads obtained, H, follows a binomial distribution where $E(H) = 100$. Find:

    a the value of p and the variance of H

    b the approximate probability of obtaining more than 110 heads.

7. One card is selected at random from each of 40 packs. Each pack contains 52 cards and includes 13 clubs. Let C be the number of clubs selected from the 40 packs.

    a Show that the variance of C is 7.5.

    b Obtain an approximation for the value of $P(C \leq 8)$, and justify the use of this approximation.

8 In a large survey, 55% of the people questioned are in full-time employment. In a random sample of 80 of these people, find:

    a the expected number in full-time employment

    b the standard deviation of the number in full-time employment

    c the approximate probability that fewer than half of the sample are in full-time employment.

9 A company manufactures rubber and plastic washers in the ratio 4:1. The washers are randomly packed into boxes of 25.

    a Find the probability that a randomly selected box contains:

       i exactly 21 rubber washers   ii exactly 10 plastic washers.

    b A retail pack contains 2000 washers. Find the expectation and variance of the number of rubber washers in a retail pack.

    c Using a suitable approximation, find the probability that a retail pack contains at most 1620 rubber washers.

10 In a certain town, 63% of homes have an internet connection.

    a In a random sample of 20 homes in this town, find the probability that:

       i exactly 15 have an internet connection

       ii exactly nine do not have an internet connection.

    b Use a suitable approximation to find the probability that more than 65% of a random sample of 600 homes in this town have an internet connection.

11 17% of the people interviewed in a survey said they watch more than two hours of TV per day. A random sample of 300 of those who were interviewed is taken. Find an approximate value for the probability that at least one-fifth of those in the sample watch more than two hours of TV per day.

12 An opinion poll was taken before an election. The table shows the percentage of voters who said they would vote for parties A, B and C.

Party	A	B	C
Votes (%)	36	41	23

Find an approximation for the probability that, in a random sample of 120 of these voters:

    a exactly 50 said they would vote for party B

    b more than 70 but fewer than 90 said they would vote for party B or party C.

13 Boxes containing 24 floor tiles are loaded into vans for distribution. In a load of 80 boxes there are, on average, three damaged floor tiles. Find, approximately, the probability that:

    a there are more than 65 damaged tiles in a load of 1600 boxes

    b in five loads, each containing 1600 boxes, exactly three loads contain more than 65 damaged tiles.

14 It is known that 2% of the cheapest memory sticks on the market are defective.

    a In a random sample of 400 of these memory sticks, find approximately the probability that at least five but at most 11 are defective.

    b Ten samples of 400 memory sticks are tested. Find an approximate value for the probability that there are fewer than 12 defective memory sticks in more than seven of the samples.

15 Randomly selected members of the public were asked whether they approved of plans to build a new sports centre and 57% said they approved. Find approximately the probability that more than 75 out of 120 people said they approved, given that at least 60 said they approved.

16 A fair coin is tossed 400 times. Given that it shows a head on more than 205 occasions, find an approximate value for the probability that it shows a head on fewer than 215 occasions.

17 An ordinary fair die is rolled 450 times. Given that a 6 is rolled on fewer than 80 occasions, find approximately the probability that a 6 is rolled on at least 70 occasions.

Checklist of learning and understanding

- A continuous random variable can take any value, possibly within a range, and those values occur by chance in a certain random manner.
- The probability distribution of a continuous random variable is represented by a function called a probability density function or PDF.
- A normally distributed random variable X is described by its mean and variance as $X \sim N(\mu, \sigma^2)$.
- The standard normal random variable is $Z \sim N(0, 1)$.
- When $X \sim N(\mu, \sigma^2)$ then $Z = \dfrac{X - \mu}{\sigma}$ has a standard normal distribution, and the standardised value $z = \dfrac{x - \mu}{\sigma}$ tells us how many standard deviations x is from the mean.
- $X \sim B(n, p)$ can be approximated by $N(\mu, \sigma^2)$, where $\mu = np$ and $\sigma^2 = npq$, provided that n is large enough to ensure that $np > 5$ and $nq > 5$.
  - $np > 5$ and $nq > 5$ are the necessary conditions for making this approximation, and larger values of np and nq result in better approximations.
- Continuity corrections must be made when a discrete distribution is approximated by a continuous distribution.

Chapter 8: The normal distribution

END-OF-CHAPTER REVIEW EXERCISE 8

1. A continuous random variable, X, has a normal distribution with mean 8 and standard deviation σ. Given that $P(X > 5) = 0.9772$, find $P(X < 9.5)$. [3]

2. The variable Y is normally distributed. Given that $10\sigma = 3\mu$ and $P(Y < 10) = 0.75$, find $P(Y \geq 6)$. [4]

3. In Scotland, in November, on average 80% of days are cloudy. Assume that the weather on any one day is independent of the weather on other days.

    i Use a normal approximation to find the probability of there being fewer than 25 cloudy days in Scotland in November (30 days). [4]

    ii Give a reason why the use of a normal approximation is justified. [1]

    Cambridge International AS & A Level Mathematics 9709 Paper 62 Q2 June 2011

4. At a store, it is known that 1 out of every 9 customers uses a gift voucher in part-payment for purchases. A randomly selected sample of 72 customers is taken. Use a suitable approximation and continuity correction to find the probability that at most 6 of these customers use a gift voucher in part-payment for their purchases. [5]

5. A survey shows that 54% of parents believe mathematics to be the most important subject that their children study. Use a suitable approximation to find the probability that at least 30 out of a sample of 50 parents believe mathematics to be the most important subject studied. [5]

6. Two normally distributed continuous random variables are X and Y. It is given that $X \sim N(1.5, 0.2^2)$ and that $Y \sim N(2.0, 0.5^2)$. On the same diagram, sketch graphs showing the probability density functions of X and of Y. Indicate the line of symmetry of each clearly labelled graph. [3]

7. The random variable X is such that $X \sim N(82, 126)$.

    i A value of X is chosen at random and rounded to the nearest whole number. Find the probability that this whole number is 84. [3]

    ii Five independent observations of X are taken. Find the probability that at most one of them is greater than 87. [4]

    iii Find the value of k such that $P(87 < X < k) = 0.3$. [5]

    Cambridge International AS & A Level Mathematics 9709 Paper 63 Q5 November 2012

8. a A petrol station finds that its daily sales, in litres, are normally distributed with mean 4520 and standard deviation 560.

    i Find on how many days of the year (365 days) the daily sales can be expected to exceed 3900 litres. [4]

    The daily sales at another petrol station are X litres, where X is normally distributed with mean m and standard deviation 560. It is given that $P(X > 8000) = 0.122$.

    ii Find the value of m. [3]

    iii Find the probability that daily sales at this petrol station exceed 8000 litres on fewer than 2 of 6 randomly chosen days. [3]

    b The random variable Y is normally distributed with mean μ and standard deviation σ. Given that $\sigma = \frac{2}{3}\mu$, find the probability that a random value of Y is less than 2μ. [3]

    Cambridge International AS & A Level Mathematics 9709 Paper 62 Q7 November 2015

9. V and W are continuous random variables. $V \sim N(9, 16)$ and $W \sim N(6, \sigma^2)$. Find the value of σ, given that $P(W < 8) = 2 \times P(V < 8)$. [4]

10 The masses, in kilograms, of 'giant Botswana cabbages' have a normal distribution with mean μ and standard deviation 0.75. It is given that 35.2% of the cabbages have a mass of less than 3 kg. Find the value of μ and the percentage of cabbages with masses of less than 3.5 kg. [5]

11 The ages of the vehicles owned by a large fleet-hire company are normally distributed with mean 43 months and standard deviation σ. The probability that a randomly chosen vehicle is more than $4\frac{1}{6}$ years old is 0.28. Find what percentage of the company's vehicles are less than two years old. [5]

12 The weights, X grams, of bars of soap are normally distributed with mean 125 grams and standard deviation 4.2 grams.

   i   Find the probability that a randomly chosen bar of soap weighs more than 128 grams. [3]

   ii Find the value of k such that $P(k < X < 128) = 0.7465$. [4]

   iii Five bars of soap are chosen at random. Find the probability that more than two of the bars each weigh more than 128 grams. [4]

Cambridge International AS & A Level Mathematics 9709 Paper 62 Q7 November 2009

13 Crates of tea should contain 200 kg, but it is known that 1 out of 45 crates, on average, is underweight. A sample of 630 crates is selected at random.

   a Find the probability that more than 12 but fewer than 17 crates are underweight. [5]

   b Given that more than 12 but fewer than 17 crates are underweight, find the probability that more than 14 crates are underweight. [5]

14 Once a week, Haziq rows his boat from the island where he lives to the mainland. The journey time, X minutes, is normally distributed with mean μ and variance σ^2.

   a Given that $P(20 \leqslant X < 30) = 0.32$ and that $P(X < 20) = 0.63$, find the values of μ and σ^2. [4]

   b The time taken for Haziq to row back home, Y minutes, is normally distributed and $P(Y < 20) = 0.6532$. Given that the variances of X and Y are equal, calculate:

      i   the mean time taken by Haziq to row back home [3]

      ii the expected number of days over a period of five years (each of 52 weeks) on which Haziq takes more than 25 minutes to row back home. [3]

15 The time taken, T seconds, to open a graphics programme on a computer is normally distributed with mean 20 and standard deviation σ.

Given that $P(T > 13 \mid T \leqslant 27) = 0.8$, find the value of:

   a σ [5]

   b k for which $P(T > k) = 0.75$. [3]

16 A law firm has found that their assistants make, on average, one error on every 36 pages that they type. A random sample of 90 typed documents, with a mean of 62 pages per document, is selected. Given that there are more than 140 typing errors in these documents, find an estimate of the probability that there are fewer than 175 typing errors. [6]

Cross-topic review exercise 3

1 a The following table shows the probability distribution for the random variable X.

x	0	1	2	3
$P(X=x)$	$\dfrac{1}{k}$	$\dfrac{3}{10}$	$\dfrac{3}{20}$	$\dfrac{1}{20}$

   i   Show that $k = 2$. [2]

   ii Calculate $E(X)$ and $Var(X)$. [3]

   iii Find the probability that two independent observations of X have a sum of less than 6. [2]

   b The following table shows the probability distribution for the random variable Y.

y	0	1	2	3
$P(Y=y)$	0.1	0.2	0.3	0.4

   If one independent observation of each random variable is made, find the probability that $X + Y = 3$. [3]

2 The random variable X has a geometric distribution such that $\dfrac{P(X=2)}{P(X=5)} = 3\tfrac{3}{8}$. Find $P(X \leqslant 3)$. [3]

3 The variable X has a normal distribution with mean μ and standard deviation σ. Given that $P(X < 32.83) = 0.834$ and that $P(X \geqslant 27.45) = 0.409$, find the value of μ and of σ. [4]

4 The length of time, in seconds, that it takes to transfer a photograph from a camera to a computer can be modelled by a normal distribution with mean 4.7 and variance 0.7225. Find the probability that a photograph can be transferred in less than 3 seconds. [3]

5 The mass of a berry from a particular type of bush is normally distributed with mean 7.08 grams and standard deviation σ. It is known that 5% of the berries have a mass of exactly 12 grams or more.

   a Find the value of σ. [3]

   b Find the proportion of berries that have a mass of between 6 and 8 grams. [3]

6 The time taken, in minutes, to fit a new windscreen to a car is normally distributed with mean μ and standard deviation 16.32. Given that three-quarters of all windscreens are fitted in less than 45 minutes, find:

   a the value of μ [3]

   b the proportion of windscreens that are fitted in 35 to 40 minutes. [3]

7 The mid-day wind speed, in knots, at a coastal resort is normally distributed with mean 12.8 and standard deviation σ.

   a Given that 15% of the recorded wind speeds are less than 10 knots, find the value of σ. [3]

   b Calculate the probability that exactly two out of 10 randomly selected recordings are less than 10 knots. [3]

   c Using a suitable approximation, calculate an estimate of the probability that at least 13 out of 100 randomly selected recordings are over 15.5 knots. [4]

8 A technical manual contains 10 pages of text, 7 pages of diagrams and 3 pages of colour illustrations. Four different pages are selected at random from the manual. Let X be the number of pages of colour illustrations selected.

   a Draw up the probability distribution table for X. [4]

   b Find:

      i $E(X)$ [2]

      ii the probability that fewer than two pages with colour illustrations are selected, given that at least one page with colour illustrations is selected. [2]

9 A fair six-sided die is numbered 1, 1, 2, 3, 5 and 8. The die is rolled twice and the two numbers obtained are added together to give the score, X.

   a Find $E(X)$. [4]

   b Given that the first number rolled is odd, find the probability that X is an even number. [2]

10 The following table shows the probability distribution table for the random variable Q.

q	1	2	3
$P(Q=q)$	$\dfrac{x-2}{x+1}$	$\dfrac{5}{18}$	$\dfrac{x-3}{x+4}$

   a Find the value of x. [3]

   b Evaluate $Var(Q)$. [2]

11 Research shows that 17% of children are absent from school on at least five days during winter because of ill health. A random sample of 55 children is taken.

   a Find the probability that exactly 10 of the children in the sample are absent from school on at least five days during winter because of ill health. [2]

   b Use a suitable approximation to find the probability that at most seven children in the sample are absent from school on at least 5 days during winter because of ill health. [4]

   c Justify the approximation made in part b. [1]

12 The ratio of adult males to adult females living in a certain town is 17 : 18, and $\dfrac{2}{9}$ of these adults, independent of gender, do not have a driving license.

   a Show that the probability that a randomly selected adult in this town is male and has a driving license is equal to $\dfrac{17}{45}$. [1]

   b Find the probability that, in a randomly selected sample of 25 adults from this town, from 8 to 10 inclusive are females who have a driving license. [4]

13 A fair eight-sided die is numbered 2, 2, 3, 3, 3, 4, 5 and 6. The die is rolled up to and including the roll on which the first 2 is obtained. Let X represent the number of times the die is rolled.

   a Find $E(X)$. [1]

   b Show that $P(X \geq 4) = \dfrac{27}{64}$. [1]

c The die is rolled up to and including the roll on which the first 2 is obtained on 20 occasions. Find, by use of a suitable approximation, the probability that $X \geq 4$ on at least half of these 20 occasions. [4]

   d Fully justify the approximation used in part **c**. [1]

14 A student wishes to approximate the distribution of $X \sim B(240, p)$ by a continuous random variable Y that has a normal distribution.

   a Find the values of p for which:

   　i approximating X by Y can be justified [3]

   　ii $\text{Var}(Y) < 45$. [3]

   b Find the range of values of p for which both the approximation is justified and $\text{Var}(Y) < 45$. [2]

PRACTICE EXAM-STYLE PAPER

Time allowed is 1 hour and 15 minutes (50 marks)

1 A mixed hockey team consists of five men and six women. The heights of individual men are denoted by h_m metres and the heights of individual women are denoted by h_w metres. It is given that $\Sigma h_w = 9.84$, $\Sigma h_m = 9.08$ and $\Sigma h_w^2 = 16.25$.

    a Calculate the mean height of the 11 team members.   [2]

    b Given that the variance of the heights of the 11 team members is $0.0416\,\text{m}^2$, evaluate Σh_m^2.   [3]

2 A and B are events such that $P(A \cap B') = 0.196$, $P(A' \cap B) = 0.286$ and $P[(A \cup B)'] = 0.364$, as shown in the Venn diagram opposite.

    a Find the value of x and state what it represents.   [2]

    b Explain how you know that events A and B are not mutually exclusive.   [1]

    c Show that events A and B are independent.   [2]

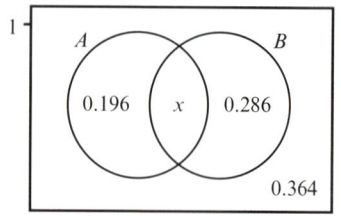

3 Meng buys a packet of nine different bracelets. She takes two for herself and then shares the remainder at random between her two best friends.

    a How many ways are there for Meng to select two bracelets?   [1]

    b If the two friends receive at least one bracelet each, find the probability that one friend receives exactly one bracelet more than the other.   [4]

4 Every Friday evening Sunil either cooks a meal for Mina or buys her a take-away meal. The probability that he buys a take-away meal is 0.24. If Sunil cooks the meal, the probability that Mina enjoys it is 0.75, and if he buys her a take-away meal, the probability that she does not enjoy it is x. This information is shown in the following tree diagram.

The probability that Sunil buys a take-away meal and Mina enjoys it is 0.156.

    a Find the value of x.   [2]

    b Given that Mina does not enjoy her Friday meal, find the probability that Sunil cooked it.   [3]

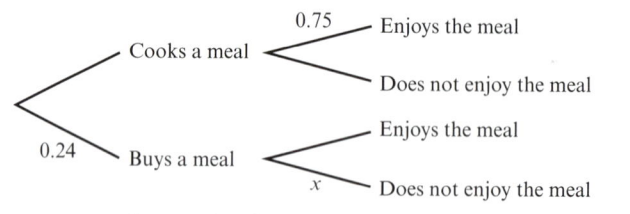

5 The following histogram summarises the total distance covered on each of 123 taxi journeys provided for customers of Jollicabs during the weekend.

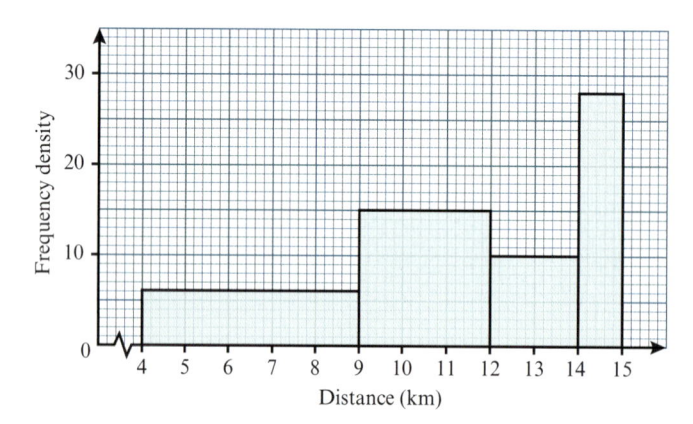

    a Find the upper boundary of the range of distances covered on these journeys.   [1]

    b Estimate the number of journeys that covered a total distance from 8 to 13 kilometres.   [2]

    c Calculate an estimate of the mean distance covered on these 123 journeys.   [3]

6   In each of a series of independent trials, a success occurs with a constant probability of 0.9.

   a The probability that none of the first n trials results in a failure is less than 0.3. Find the least possible value of n. [2]

   b State the most likely trial in which the first success will occur. [1]

   c Use a suitable approximation to calculate an estimate of the probability that fewer than 70 successes occur in 80 trials. [4]

7   The following stem-and-leaf diagram shows the number of shots taken by 10 players to complete a round of golf.

$$\begin{array}{c|cccc}
6 & 1 & y & & \\
7 & 0 & 1 & 2 & x \\
8 & 0 & 1 & 9 & 9
\end{array}$$

Key: $6\,|\,1$ represents 61 shots

   a Given that the median number of shots is 74.5 and that the mean number of shots is 75.4, find the value of x and of y. [3]

The numbers of golf shots are summarised in a box-and-whisker diagram, as shown.

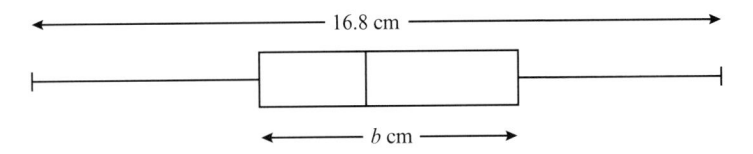

   b Given that the whisker is 16.8 cm long, find the value of b, if the width of the box is b cm. [3]

   c Explain why the mode would be the least appropriate measure of central tendency to use as the average value for this set of data. [1]

8   To conduct an experiment, a student must fit three capacitors into a circuit. He has eight to choose from but, unknown to him, two are damaged. He fits three randomly selected capacitors into the circuit. The random variable X is the number of damaged capacitors in the circuit.

   a Draw up the probability distribution table for X. [3]

   b Calculate $\text{Var}(X)$. [3]

   c The student discovers that exactly one of the capacitors in the circuit is damaged but he does not know which one. He removes one capacitor from the circuit and replaces it with one from the box, both selected at random. Find the probability that the circuit now has at least one damaged capacitor in it. [4]

THE STANDARD NORMAL DISTRIBUTION FUNCTION

If Z is normally distributed with mean 0 and variance 1, the table gives the value of $\Phi(z)$ for each value of z, where

$$\Phi(z) = P(Z \leq z).$$

Use $\Phi(-z) = 1 - \Phi(z)$ for negative values of z.

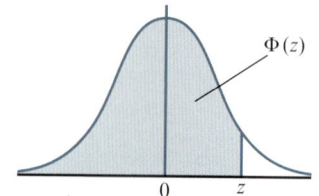

z	0	1	2	3	4	5	6	7	8	9	1	2	3	4	5	6	7	8	9
														ADD					
0.0	0.5000	0.5040	0.5080	0.5120	0.5160	0.5199	0.5239	0.5279	0.5319	0.5359	4	8	12	16	20	24	28	32	36
0.1	0.5398	0.5438	0.5478	0.5517	0.5557	0.5596	0.5636	0.5675	0.5714	0.5753	4	8	12	16	20	24	28	32	36
0.2	0.5793	0.5832	0.5871	0.5910	0.5948	0.5987	0.6026	0.6064	0.6103	0.6141	4	8	12	15	19	23	27	31	35
0.3	0.6179	0.6217	0.6255	0.6293	0.6331	0.6368	0.6406	0.6443	0.6480	0.6517	4	7	11	15	19	22	26	30	34
0.4	0.6554	0.6591	0.6628	0.6664	0.6700	0.6736	0.6772	0.6808	0.6844	0.6879	4	7	11	14	18	22	25	29	32
0.5	0.6915	0.6950	0.6985	0.7019	0.7054	0.7088	0.7123	0.7157	0.7190	0.7224	3	7	10	14	17	20	24	27	31
0.6	0.7257	0.7291	0.7324	0.7357	0.7389	0.7422	0.7454	0.7486	0.7517	0.7549	3	7	10	13	16	19	23	26	29
0.7	0.7580	0.7611	0.7642	0.7673	0.7704	0.7734	0.7764	0.7794	0.7823	0.7852	3	6	9	12	15	18	21	24	27
0.8	0.7881	0.7910	0.7939	0.7967	0.7995	0.8023	0.8051	0.8078	0.8106	0.8133	3	5	8	11	14	16	19	22	25
0.9	0.8159	0.8186	0.8212	0.8238	0.8264	0.8289	0.8315	0.8340	0.8365	0.8389	3	5	8	10	13	15	18	20	23
1.0	0.8413	0.8438	0.8461	0.8485	0.8508	0.8531	0.8554	0.8577	0.8599	0.8621	2	5	7	9	12	14	16	19	21
1.1	0.8643	0.8665	0.8686	0.8708	0.8729	0.8749	0.8770	0.8790	0.8810	0.8830	2	4	6	8	10	12	14	16	18
1.2	0.8849	0.8869	0.8888	0.8907	0.8925	0.8944	0.8962	0.8980	0.8997	0.9015	2	4	6	7	9	11	13	15	17
1.3	0.9032	0.9049	0.9066	0.9082	0.9099	0.9115	0.9131	0.9147	0.9162	0.9177	2	3	5	6	8	10	11	13	14
1.4	0.9192	0.9207	0.9222	0.9236	0.9251	0.9265	0.9279	0.9292	0.9306	0.9319	1	3	4	6	7	8	10	11	13
1.5	0.9332	0.9345	0.9357	0.9370	0.9382	0.9394	0.9406	0.9418	0.9429	0.9441	1	2	4	5	6	7	8	10	11
1.6	0.9452	0.9463	0.9474	0.9484	0.9495	0.9505	0.9515	0.9525	0.9535	0.9545	1	2	3	4	5	6	7	8	9
1.7	0.9554	0.9564	0.9573	0.9582	0.9591	0.9599	0.9608	0.9616	0.9625	0.9633	1	2	3	4	4	5	6	7	8
1.8	0.9641	0.9649	0.9656	0.9664	0.9671	0.9678	0.9686	0.9693	0.9699	0.9706	1	1	2	3	4	4	5	6	6
1.9	0.9713	0.9719	0.9726	0.9732	0.9738	0.9744	0.9750	0.9756	0.9761	0.9767	1	1	2	2	3	4	4	5	5
2.0	0.9772	0.9778	0.9783	0.9788	0.9793	0.9798	0.9803	0.9808	0.9812	0.9817	0	1	1	2	2	3	3	4	4
2.1	0.9821	0.9826	0.9830	0.9834	0.9838	0.9842	0.9846	0.9850	0.9854	0.9857	0	1	1	2	2	2	3	3	4
2.2	0.9861	0.9864	0.9868	0.9871	0.9875	0.9878	0.9881	0.9884	0.9887	0.9890	0	1	1	1	2	2	2	3	3
2.3	0.9893	0.9896	0.9898	0.9901	0.9904	0.9906	0.9909	0.9911	0.9913	0.9916	0	1	1	1	1	2	2	2	2
2.4	0.9918	0.9920	0.9922	0.9925	0.9927	0.9929	0.9931	0.9932	0.9934	0.9936	0	0	1	1	1	1	1	2	2
2.5	0.9938	0.9940	0.9941	0.9943	0.9945	0.9946	0.9948	0.9949	0.9951	0.9952	0	0	0	1	1	1	1	1	1
2.6	0.9953	0.9955	0.9956	0.9957	0.9959	0.9960	0.9961	0.9962	0.9963	0.9964	0	0	0	0	1	1	1	1	1
2.7	0.9965	0.9966	0.9967	0.9968	0.9969	0.9970	0.9971	0.9972	0.9973	0.9974	0	0	0	0	0	1	1	1	1
2.8	0.9974	0.9975	0.9976	0.9977	0.9977	0.9978	0.9979	0.9979	0.9980	0.9981	0	0	0	0	0	0	0	1	1
2.9	0.9981	0.9982	0.9982	0.9983	0.9984	0.9984	0.9985	0.9985	0.9986	0.9986	0	0	0	0	0	0	0	0	0

Critical values for the normal distribution

The table gives the value of z such that $P(Z \leq z) = p$, where $Z \sim N(0, 1)$.

p	0.75	0.90	0.95	0.975	0.99	0.995	0.9975	0.999	0.9995
z	0.674	1.282	1.645	1.960	2.326	2.576	2.807	3.090	3.291

Answers

1 Representation of data

Prerequisite knowledge
1. **a** 62 m **b** 256.25 m²
2. Frequencies are equal because areas are equal.
3. **a** 27 **b** 6

Exercise 1A
1.
   ```
   0 | 1 2 3 3 4 4 5 6 7 8 9    Key: 1 | 0
   1 | 0 1 2 3 3 5 6              represents 10
   2 | 0 6                        visits
   ```
2. **a**
   ```
   15 | 0 2 6 8 9    Key: 15 | 0
   16 | 0 2 3 5      represents
   17 | 0 2 5        150 coins
   ```
   b $1615
3. **a** 18 **b** 8 **c** 20%
   d i 30–39 **ii** 10–19
4. **a** 88 **b** $10.80 **c** 0 and 3
5. **a**
Batsman P		Batsman Q	
	2	0 1	Key: 6 | 3 | 1
9 8 7 7 6	3	1 6	represents 36
8 7 4 1 1	4	2 5 8	runs for P and
9 9 7 3 2	5	1 2 6 7	31 runs for Q
	6	4 8	
	7	1 7	

   b i Q; scored more runs.
   ii P; scores are less spread out.
6. **a**
Wrens (10)		Dunnocks (10)	
	3	1	Key: 8 | 1 | 9
9 8 7	1	7 9	represents 18 eggs
4 3 3 2 1 0	2	2 2 3 4	for a wren and 19
	2	5 7 8	eggs for a dunnock
	3	0	

   b 218 **c** 93%
7. The girl who scored 92%; 5 boys.

Exercise 1B
1. **a** 175 and 325 years
   b All 150 years
   c Boundaries at 25, 175, 325, 475, 625 years.
   Densities ∝ 15, 18, 12, 6 (such as 0.1, 0.12, 0.08, 0.04).
   d 15
2. **a** 70
   b Boundaries at 4, 12, 24, 28 grams.
   Densities ∝ 28, 33, 17.5.
   c 310
3. **a** 33
   b Boundaries at 1.2, 1.3, 1.6, 1.8, 1.9 m.
   Densities ∝ 170, 110, 210, 80.
   c 29
4. **a** Any u from 35 to 50.
   b Boundaries at 0, 5, 15, 30, u cm.
   Densities ∝ 12.8, 23.2, 16, $\frac{64}{u-30}$.
   c i 456 **ii** 246
5. **a** 2.85 − 2.55 = 0.3
   b Boundaries at 2.55, 2.85, 3.05, 3.25, 3.75 min.
   Densities ∝ 50, 125, 100, 20.
   c 2 min 45 s or 165 s.
   d i 3.5 **ii** 3.01
6. **a** 324
   b i 30 **ii** 92
   c Proof
   d 440; Population and sample proportions are the same.
7. **a** 480 **b** 130 **c** 110
8. **a** $\frac{17}{23}$ **b** 399 **c** 12.6 cm
9. **a** 12 : 8 : 3 **b** $n = \frac{1}{150}$
   c i 210 **ii** 36
   d $0.215 \leq k < 0.720$ mm
   We can be certain only that $0.1 \leq a < 0.4$ and that $0.4 \leq b < 0.8$.
10. **a** $a = 159, b = 636$ **b** 23.5 kg
11. $\frac{4hd}{5n}$
12. 33 cm
13. $p = 29, q = 94$

Exercise 1C
1. **a** Points plotted at (1.5, 0), (3, 3), (4.5, 8), (6.5, 32), (8.5, 54), (11, 62), (13, 66).
   b i 23 **ii** 7.8 s
2. **a** 19.5 cm
   b
Width (cm)	<9.5	<14.5	<19.5	<29.5	<39.5	<44.5
No. books (cf)	0	3	16	41	65	70

Points plotted at (9.5, 0), (14.5, 3), (19.5, 16), (29.5, 41), (39.5, 65), (44.5, 70).

    c i 34 or 35   ii ≈ 33.25 to 44.5 cm

3 a Points plotted at (0.10, 0), (0.35, 16), (0.60, 84), (0.85, 144), (1.20, 156) for A.

      Points plotted at (0.10, 0), (0.35, 8), (0.60, 62), (0.85, 120), (1.20, 156) for B.

  b i ≈ 107 for engine A; ≈ 87 for engine B.

    ii ≈ 108

  c ≈ 42

4 a 17; *cf*s 20 and 37 are precise.

  b i 12   ii 28

  c $k = 4.7$ to 4.8

  d It has the highest frequency density.

5 a i 64   ii 76

  b ≈ 7.4 g

  c (12, 304)

6 $a = 32, b = 45, c = 15, d = 33$

7 a 65   b 24

8 a Ratio of under 155 cm to over 155 cm is 3:1 for boys and 1:3 for girls.

  b 81 or 82

  c There are equal numbers of boys and girls below and above this height.

  d Polygon or curve through (140, 0), (155, 25), (175, 50).

9 a Points plotted at (18, 0), (20, 27), (22, 78), (25, 89), (29, 94), (36, 98), (45, 100).

  b 27 years and 4 or 5 months

  c i 1000

    ii All age groups are equally likely to find employment.

      Either with valid reasoning; e.g. underestimate because older graduates with work experience are more attractive to employers.

10 Points plotted at (4.4, 0), (6.6, 5), (8.8, 12), (12.1, 64), (15.4, 76), (18.7, 80) for new cars.

    Points plotted at (4, 0), (6, 5), (8, 12), (11, 64), (14, 76), (17, 80) for ⩾ 100 000 km.

    Polygons 17 cars; curves ≈ 16 cars.

11 a Points plotted at (1.0, 0), (1.5, 60), (2.0, 182), (2.5, 222), (3.0, 242) for diameters.

    Points plotted at (2.0, 0), (2.5, 8), (3.0, 40), (3.5, 110), (4.0, 216), (4.5, 242) for lengths.

  b Least $n = 0$; greatest $n = 28$.

  c Diameter and length for individual pegs are not shown.

    Best estimate is 'between 171 and 198 inclusive'.

    The length and diameter of each peg should be recorded together, then the company can decide whether each is acceptable or not.

Exercise 1D

1 a Any suitable for qualitative data.

  b Pie chart, as $\frac{3}{4}$ circle easily recognised, or a sectional percentage bar chart.

2 Histogram; area of middle three columns > half total column area.

3 a Numbers can be shown in compact form on three rows; bar chart requires 17 bars, all with frequencies 0 or 1.

  b Sum = 100 shows that 11 boxes of 100 tiles could be offered for sale.

4 a 7 months

  b Percentage *cf* graph; passes below the point (12, 100).

5 a Histogram: Frequency density may be mistaken for frequency.

    Pie chart: does not show numbers of trees.

  b Pictogram: short, medium, tall; two, three and four symbols, each for six trees, plus a key.

    Shows 12, 18, 24 and a total of 54 trees.

6 a

1. Score (%)	30–39	40–49	50–59	60–69	70–79	80–89	90–99
Frequency	3	5	6	15	5	4	2

  b

2. Grade	C	B	A
Frequency	8	26	6

Any three valid, non-zero frequencies that sum to 40.

c Raw: stem-and-leaf diagram is appropriate.

  Tables 1 and 2 do not show raw marks, so these diagrams are not appropriate.

  Table 1: Any suitable for grouped discrete data; e.g. histogram.

  Table 2: Any suitable for qualitative data.

7 a E.g. He worked for less than 34 hours in 49 weeks, and for more than 34 hours in 3 weeks.

  b It may appear that Tom worked for more than 34 hours in a significant number of weeks.

  c Histogram: boundaries at 9, 34 and 44; densities ∝ 98 and 15.

  Pie chart: sector angles ≈ 339.2° and 20.8°.

  Bar chart: frequencies 49 and 3.

  Sectional percentage bar chart: ≈ 94.2 and 5.8%.

8 a Some classes overlap (are not continuous).

  b Refer to focal lengths as, say, A to E in a key.

  Pie chart: sector angles 77.1°, 128.6°, 77.1°, 51.4°, 25.7°.

  Bar chart or vertical line graph: heights 18, 30, 18, 12, 6.

  Pictogram: symbol for 1, 3 or 6 lenses.

9

Country	C	SL	Ma	G	Mo
% of population	14.4	8.9	3.8	17.7	27.4

End-of-chapter review exercise 1

1 i 50

  ii Boundaries at 20, 30, 40, 45, 50, 60, 70 g.

  Frequency densities ∝ 2, 3, 10, 12, 5, 1.

2 16.5, 3 and 18 cm

3 a 6

  b Quantitative and continuous

4 a 6

  b Five additional rows for classes 0–4, 20–24, 25–29, 30–34, 35–39.

5 $a = 9, b = 2$

6 a 48

  b 0.7 cm

7 a 120, 180 and 90

  b 6.75 cm

  c There is a class between them (not continuous).

8 a 30 days for region A, 31 days for region B.

  b Bindu: unlikely to be true but we cannot tell, as the amount of sunshine on any particular day is not shown.

  Janet: true (max. for region A is 106 h; min. for region B is 138 h).

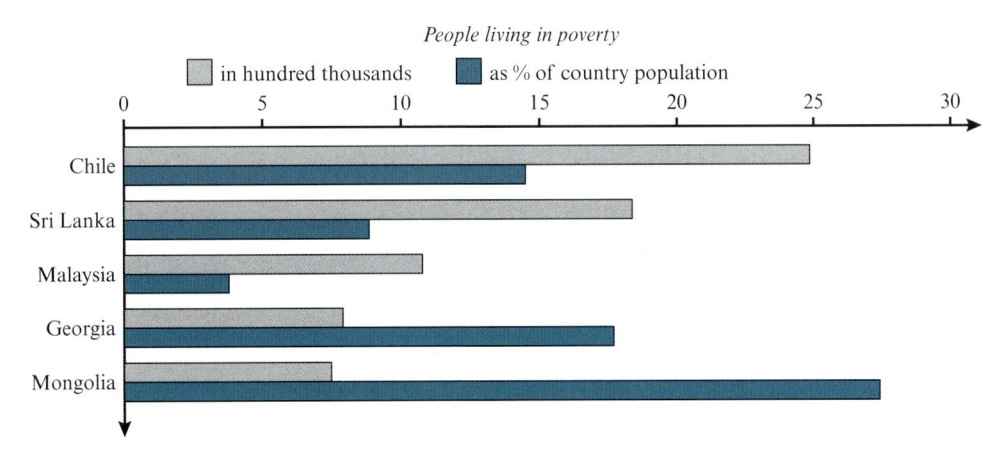

People living in poverty — in hundred thousands / as % of country population

Chile, Sri Lanka, Malaysia, Georgia, Mongolia

Mongolia, for example, has the lowest number, but the highest percentage, of people living in poverty.

9 i Points plotted at:
   (20.5, 10), (40.5, 42), (50.5, 104),
   (60.5, 154), (70.5, 182), (90.5, 200) or
   (20, 10), (40, 42), (50, 104),
   (60, 154), (70, 182), (90, 200) or
   (21, 10), (41, 42), (51, 104),
   (61, 154), (71, 182), (91, 200).
   ii 174 to 180
   iii 58, 59 or 60

2 Measures of central tendency

Prerequisite knowledge
1 Mean = 5, median = 4.3, mode = 3.9.
2 1.94

Exercise 2A
1 a No mode   b 16, 19 and 21
2 'The' is the mode.   3 7 for x; −2 for y.
4 14–20 for x; 3–6 for y.
5 Most popular size(s) can be pre-cut to serve customers quickly, which may result in less wastage of materials.
6 216   7 69   8 73

Exercise 2B
1 a 50   b 7.1   c $4\frac{13}{40}$
2 a $p = \pm 7$ b $q = 9$ or −10
3 a 23.25   b 1062   c 88
   d 12   e 113.67
4 a 19   b 3.68825
5 $a = 12$
6 a 4.1   b 24.925
7 73.8%   8 $1846
9 30 years; the given means may only be accurate to the nearest month.
   Actual age could be any from 28 yr 8.5 m to 31 yr 3.5 m.
10 a Mean ($10) is not a good average; 36 of the 37 employees earn less than this.
   b $7.25
11 a $143 282
   41% means from 1495 to 1531 passengers.
   29% means from 1052 to 1088 passengers.
   b $k = 252$ or $k = 396$, depending on assumptions

12 $9\frac{2}{9}$ cm
13 a 54.6   b 59.0
    c The scales may have underestimated or overestimated masses. Not all tomatoes may have been sold (i.e. some damaged and not arrived at market).
14 a 1.5
    b i 1.96   ii 3.48
    c For example, bar chart with four groups of four bars, or separate tables for boys and girls.
15 $n = 12$
    None of the 120 refrigerators have been removed from the warehouse.
16 One more day required.
    He works at the same rate or remaining rooms take a similar amount of time (are of a similar size).
17 a i 5.89 cm   ii 5.76 cm
    b 152.0°

Exercise 2C
1 a 74   b 94   c 64
2 18   3 204   4 40.35 mm
5 −0.8
6 a To show whether the cards fit ($x < 0$), or not ($x > 0$).
   b 2%
   c $-0.0535 + 24 = 23.9465$ mm
7 Fidel; Fidel's deviations > 0, Ramon's deviations < 0.
8 63.5 s; accurate to 1 decimal place.
9 a $3n$; 60°   b $90n$
10 3.48   11 $1.19

Exercise 2D
1 5700; the total mass of the objects, in grams.
2 a $\Sigma 5x$ or $5\Sigma x$
   b $\Sigma 0.001x$ or $0.001\Sigma x$
3 $\Sigma 0.01w$ or $0.01\Sigma w$
4 3.6
5 a Calculate estimate in mph, then multiply by $\frac{8}{5}$ or 1.6.
   b $19.7625 \times 1.6 = 31.62$ km/h

6 $\bar{x} = 12.4$; $b = -3$
7 $a = 4, b = 5$
8 a (−1.8, 2.8)   b (26, −6)
   c TE(5.2, −1.2) → (19, −2)
     ET(5.2, −1.2) → (−9, 14)
     Location is dependent on order of transformations.
9 $p = 40$; $q = 12\,000$; $\$75\,000$
   Appears unfair; the smaller the amount invested, the higher the percentage profit.
10 $281\bar{x}$ g/cm^2

Exercise 2E

1 a 15
   b Median; it is greater than the mean (12.4).
   c For example, being unable to pay a bill because of low earnings.
2 a 11.5
   b Negatively skewed; $\bar{t} = 10.9 <$ median.
3 a Median = 6; mode = 8
   b Median is central to the values but occurs less frequently than all others.
     Mode is the most frequently occurring value but is also the highest value.
   c Two incorrect
4 a ≈ 4.4 min
   b 2.8 and 6.4 min
5 Points plotted at (0, 0), (0.2, 16), (0.3, 28), (0.5, 120), (0.7, 144), (0.8, 148).
   Median = 0.4 kg
   a 92   b 32
6 Mode (15) and median (16) unaffected; mean decreases from 16 to 14.75.
7 a Points plotted at (85, 0), (105, 12), (125, 40), (145, 94), (165, 157), (195, 198), (225, 214), (265, 220).
     Polygon and curve give median ≈ 150 days.
   b Likely to use whichever is the greatest. Estimate of mean (152.84) appears advantageous. (They could consider using the greatest possible mean, 164.41.)

8 'Average' could refer to the mean, the median or the mode.
   Median > 150; Mean < 150.
   150 is close to lower boundary of modal class.
   Claim can be neither supported nor refuted.
9 There is no mode.
   Mean ($1 049 500) is distorted by the expensive home.
   Median ($239 000) is the most useful.
10 a $p = 12, q = 40, r = 54$
   b i Reflection in a horizontal line through cf value of 30.
     ii Median safe current = median unsafe current
11 a First-half median is in 1–2; second-half median is in 4–5.
   b i 3
     ii First half data are positively skewed (least possible mean is 100.8 s).
12 a Points plotted at (0, 0), (26, 15), (36, 35), (50, 60), (64, 75), (80, 80).
     Median = 39%
     New points at (0, 0), (16, 9.23), (26, 15), (40, 42.1), (54, 64.2), (80, 80).
   b 20
13 a i
     ii Mode = mean = median = 8.
   b No effect on mode or median. Mean increases to 9. Curve positively skewed.
   c $b = -11$; No effect on mode or median. Curve negatively skewed.
14 Any symmetrical curve with any number of modes (or uniform).
15 a Symmetrical; mean = median = mode
   b Chemistry: negatively skewed; Physics: positively skewed.

End-of-chapter review exercise 2

1. a Mean < median and mode.
   b Mean > median and mode.
   c Mean = median and mode.
2. $n = 22; 623\,g$
3. a Mode = 13
   b Median = 28
   c 55
4. a 15.15
   b 13.3
5. a 6
   b i 14
      ii 25
6. 25
7. a Proof
   b $0.18; it is an estimate of the mean amount paid.
8. a Mode indicates the most common response.
      Median indicates a central response (one of the options or half-way between a pair).
   b Allows for a mean response, which indicates which option the average is closest to.
9. a 1
   b No; it is the smallest value and not at all central.
   c 11
   d Positively skewed; mode < median < mean.
10. i Boundaries at 0.05, 0.55, 1.05, 2.05, 3.05, 4.55 h.
       Frequency densities $\propto$ 22, 30, 18, 30, 14.
    ii 2.1 h
11. 16.4
12. 81
13. a Mode = 0, mean = 1, median = 0
    b Mean; others might suggest that none of the items are damaged.
14. a 4006 − 2980 = $1026   b $3664
15. 1.95

3 Measures of variation

Prerequisite knowledge

1. 16 cm
2. a 4.5   b 27.3

Exercise 3A

Box plots given by: smallest ... Q_1 ... Q_2 ... Q_3 ... largest / Item (units), as appropriate.

1. a 25 and 17   b 35 and 20
   c 65 and 25   d 96 and 59
   e 8.5 and 5.6
2. a Range = 3.3; IQR = 1.75
   b Negative
3. a 41 and 18
   b 9 ... 28 ... 37 ... 46 ... 50 / Marks.
   c $Q_3 = 2Q_2 - Q_1$
4. a Yes, if the range alone is considered.
   b Hockey: 11 ... 13 ... 17 ... 20 ... 24 / Fouls.
      Football: 10 ... 18.5 ... 20 ... 22.5 ... 23 / Fouls.
      with the same scale.
      Fewer fouls on average in hockey but the numbers varied more than in football.
5. a Ranges and IQRs are the same (35 and 18) but their marks are quite different.
   b One of median (33/72) or mean (33/72) and one of range or IQR.
6. a Points plotted at (35, 0), (40, 20), (45, 85), (50, 195), (55, 222), (70, 235), (75, 240).
      35 ... 43.1 ... 46.6 ... 49.3 ... 75 / Speed (km/h), parallel to speed axis.
   b Positive skew
7. a i Males: 0 ... 0 ... 3 ... 14 ... 39 / Trips abroad.
         Females: 3 ... 5 ... 12 ... 20 ... 22 / Trips abroad.
         Same scale.
      ii Males: range = 39; IQR = 14; median = 3
          Females: range = 19; IQR = 15; median = 12
          On average, females made more trips abroad than males. Excluding the male who made 39 trips, variation for males and females is similar.
   b No, there are no data on the number of different countries visited.
8. a $\approx 0.130\,\Omega$ b $\approx 0.345\,\Omega$
   c $\approx$ 68th percentile   d $\approx 0.095\,\Omega$
9. a 52 cm^2
   b 4.0 ... 25.8 ... 33.2 ... 38.8 ... 56.0 / Area (cm^2).
   c 15.2 to 16.0 cm^2
   d Area < 6.3 cm^2 or area > 58.3 cm^2.
      Estimate $\approx$ 8 (any from 0 to 15)

10 a Points plotted at (–1.5, 0), (–1.0, 24), (–0.5, 70), (0, 131), (0.5, 165), (1.0, 199), (1.5, 219), (2.5, 236).

  b 89.9° and 1.3°

  c 18%

11 a 10   b 30

12 a Points plotted at (0, 0), (4, 2), (11, 21), (17, 44), (20, 47), (30, 50).

  b i ≈ 0.06 g

   ii ≈ 0.12 g

  c n ≈ 40

  d Variation is quite dramatic (from 0 up to a possible 3% of mass).

   Mushrooms are notoriously difficult to identify (samples may not all be of the same type). Toxicity varies by season.

13 Should compare averages and variation (and skewness) and assess effectiveness in reducing pollution level for health benefits.

Exercise 3B

1 a Mean = 37.5, SD = 12.4
  b Mean = 0.45, SD = 9.23
2 a Var(B) = Var(C) = Var(P) = 96
  b The three values are identical.
   No; mean marks are not identical (B = 33, C = 53 and P = 63).
3 Mean = $1\frac{24}{35}$ or 1.69; variance = 1.64
4 a Mean = 2; SD = 0.803
  b $Q_1 = Q_3 = 2$, so IQR = 0.
   That the middle 50% of the values are identical.
5 a Girls: mean = 40, SD = 13.0 min
   Boys: mean = 40, SD = 16.3 min
  b i On average, the times spent were very similar.
   ii Times spent by boys are more varied than times spent by girls.
6 5.94 cm
7 k = 6; Var(x) = 2.72
8 a a = 13, b = 40   b 6.23 cm
9 k = 43; SD = 12.5 km; IQR = 24 km; IQR ≈ 2 × SD
10 a Mean = 0.97 t; SD = 0.44 t
  b Mean decreases to 0.73 t; SD increases to 0.57 t.

11 x = 12, y = 18
   Gudrun is 22 years old.
   Variance increases from 69.12 to 72.88 years^2.
   None of the original 50 staff have been replaced.
12 a Mean decreases by 11.6 cm.
    Median decreases by 40.3 cm.
  b SD increases by 116 cm.
    IQR increases by 216 cm.
    (Range increases by 344 cm.)
  c Discs get closer to P, but distances become more varied.
  d Proof

Exercise 3C

1 a 65.375   b 9.17
  c 120   d 161 800
  e 28
2 $n = 20; \bar{x} = 11$
3 2.15
4 Mean = 60.2 kg; SD = 14.1 kg
5 a Proof   b 27.2 psi
6 a $\sum y^2 = \dfrac{52n^2 - 4915n + 616549}{n + 29}$
  b n = 35
7 687.5   8 31
9 Proof
10 $\dfrac{S}{\sqrt{2}}$ or $\dfrac{S\sqrt{2}}{2}$.

Exercise 3D

1 Men 8 kg; women 6 kg
2 1.5
3 7.92 mm and 24 009.8
4 8
5 n = 15
6 Mean is not valid (it is 165 cm); standard deviation is valid.
7 Mean = 4 h 20 min; SD = 7.3 min
  If 10-minute departure delay avoids busy traffic conditions.
8 0.96 cm^2

9　a　Mean = 8; SD = 4　　b　Mean = 7; SD = 4

　　c　$\frac{n^2 - 1}{3}$; variance of the first n positive odd integers.

10　a　43　　b　Proof　　c　1.179

11　a　Proof

　　b　$\sum y = 1104, \sum y^2 = 15\,416$　c　20.1376

12　a　162.14 cm.

　　b　($\sum x^2 = 5\,720\,640, \sum y^2 = 7\,445\,100$);

　　　Var(X) = 42.1004 cm^2

Exercise 3E

1　$0.64　　2　8.5　　3　2.64

4　a　133 and 2673

　　b　0.457 °C, using 133 and 2673.

　　c　0.209 (°C)^2

5　$75 600

6　a　27°F

　　b　Mean = 12.5 °C; SD = 4.5 °C

7　a　Fruit & veg; mean unchanged, so total unchanged.

　　b　Tinned food; mean increased but standard deviation unchanged.

　　c　Bakery; mean and standard deviation decreased by 10%.

8　26 m　　　　　9　18.0% increase

End-of-chapter review exercise 3

1　a　Proof

　　b　$0.917 or $0.92

2　a　0

　　b　0, 1, 2, 3 or 4

3　Five

4　a　0.319 m

　　b　Mean increased by 1.5 cm (to 90 cm); SD unchanged.

5　a　Marks are improving and becoming more varied.

　　b　Third test

　　c　First test positive; second test negative

6　a　97.92 cm

　　b　11.5 cm

7　a　Range = 139; IQR = 8; SD = 37.7

　　b　IQR; unaffected by extreme value (180).

8　i　173 cm

　　ii　834 728.6 and 4.16 cm

9　45.8 and 14.9 s

10　i

Squad A		Squad B
	7	5 7 9
4 4 2	8	2 3 4 6
9 8 7 6 1	9	4 5 6
9 7 4 0	10	1 8
6 5	11	1 3 5
2	12	

Key: 1 | 9 | 4 represents 91 kg for squad A and 94 kg for squad B

　　ii　18 kg

　　iii　103.4 kg

11　i　126.5 cm

　　ii　4908.52 cm^2

12　i　Mean = 40.9 or $40\frac{8}{9}$; SD = 8.30

　　ii　8.41

13　5514

14　SD increases by 68.4%.

　　IQR increases by 9.30% (or 6.90%, depending on method).

　　Proportional change in SD is much greater than in IQR.

15　14.0 cm

16　a　SD = 21.5

　　b　Mean = −2; SD = 21.5

　　　Mean is affected by addition of −202 but SD is unaffected.

17　a　5.6 > 2 × 2.75　b　803 ≠ 140^2; 132 ≠ 44^2　c　1.63

18　a　19.8 − 18 = 1.8

　　b　$\sum a^2 = 1964.46, \sum b^2 = 2278.12$; 1.99 years^2

Cross-topic review exercise 1

1　a　25

　　b　Player A = 25, player B = 21

　　c　
```
1 | 8 9
2 | 0 1 1 2 2 3 4
2 | 5 6 7 8
3 | 3
```
Key: 1 | 8 represents 18 games

2　a　10–15 and 26–30

　　　26 − 15 = 11

　　b　

No. incorrect answers	0	1–9	10–14	15–24	25–30	31–40
No. candidates	1	19	23	27	24	18

　　c　19.6

3　a　2.3 cm

　　b　0.0178 m

4   a   17.5
    b   126.3125
    c   Student A and student F
5   Higher average and less varied growth.
6   a
```
0 | 8 9
1 | 1 3 3 4 5 7 7 8 8      Key: 1 | 1
2 | 0 1 2 5 6 6 7 9        represents 11
3 | 1 2 5 6                unwanted emails
```
    b   8 ... 14 ... 20 ... 27 ... 36 / Unwanted emails.
7   a   5.94 and 6.685
    b   Mean = 2 990 000 or 2.99×10^6
        SD = 366 151 or 3.66151×10^5
8   a   32
    b   70 and 75 km/h
    c   72.3 km/h
9   a   Proof
    b   36.09 g and 0.67 g
    c $0.4489 g^2$
10 a   i   75
        ii ≈ 69
    b   12
11 a   50 and 1.80
    b $C = 50 - S$
        SD(C) = SD(aS + b)
        a = −1 and b = 50
12 a   1.48
    b   79
    c $a = 7, b = 9$
13 i   Median = 0.825 cm; IQR = 0.019 cm
    ii $q = 4, r = 2$
    iii X: 0.802 ... 0.814 ... 0.825 ... 0.833 ... 0.848 / Length (cm)
        Y: 0.811 ... 0.824 ... 0.837 ... 0.852 ... 0.869 / Length (cm)
        Same scale
    iv Longer on average in Y; less varied in X.

4 Probability

Prerequisite knowledge
1   30
2   $\dfrac{1}{12}$

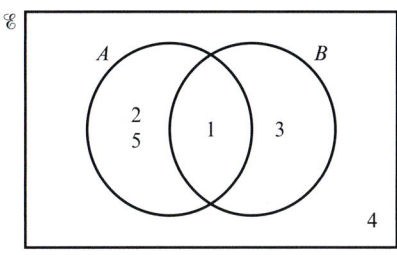

3
n(A ∪ B′) = 4 and n(A′ ∩ B) = 1

Exercise 4A
1   a $\dfrac{1}{36}$ b $\dfrac{2}{3}$
2   a   The team's previous results.
    b   8
    c   They may win some of the games that they are expected to draw.
3   12
4   a   300   b   At least 240
5   a   5   b   15
6   50
7   $\dfrac{3}{8}$
8   $\dfrac{1}{1953}$

Exercise 4B
1   a $\dfrac{2}{3}$ b $\dfrac{2}{3}$ c $\dfrac{5}{6}$
2   a   Girls who took the test.
    b $\dfrac{23}{40}$
3   a   i $\dfrac{3}{5}$ ii $\dfrac{10}{11}$
    b   Not a female sheep. Not a male goat.
4   a   i   (3, 3)
        ii (2, 4) and (4, 2)
        iii (2, 2), (4, 4), (6, 6)
    b   X, Y and Z are not mutually exclusive.
5   a $\dfrac{1}{2}$ b $\dfrac{7}{8}$
6   a $a = 7, b = 2, c = 6$
    b   i $\dfrac{3}{5}$ ii $\dfrac{13}{25}$

7 a

[Venn diagram: C and H intersecting; C only = 12, intersection = 7, H only = 13, outside = 8]

b i $\frac{3}{10}$ ii $\frac{5}{8}$

8 44%

9 a $\frac{10}{11}$ b $\frac{6}{11}$ c $\frac{7}{22}$

10 a Students who study Pure Mathematics and Statistics but not Mechanics.
   b i $\frac{89}{100}$ ii $\frac{6}{25}$
   c Mechanics, Statistics, Pure Mathematics

11 a No; $P(X \cap Y) \neq 0$ or equivalent.
   b 0.9 c 0.7

12 a A and C b 0.22

13 a $\frac{2}{75}$ b $\frac{31}{75}$ c $\frac{29}{75}$

14 a 0.6 b 0.4

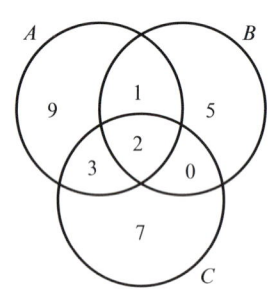

15 a

[Venn diagram with three circles A, B, C: A only = 9, B only = 5, C only = 7, A∩B only = 1, A∩C only = 3, B∩C only = 0, A∩B∩C = 2]

b 19; they had not visited Burundi.
c They had visited Angola or Burundi but not Cameroon; 15.
d $\frac{2}{9}$

Exercise 4C

1 $\frac{1}{2}$

2 a $\frac{1}{36}$ b $\frac{1}{4}$ c $\frac{1}{9}$

3 a 0.012 b 0.782

4 0.42

5 a 0.84 b 0.85

6 a 0.343 b 0.441

7 a i 0.544 ii 0.3264 iii 0.4872
  b The result in any event has no effect on probabilities in other events.
    E.g. winning one event may increase an athlete's confidence in others.

8 a $\frac{5}{24}$ b $\frac{3}{8}$

9 a Untrue. Any number from 0 to 10 may be delivered; 9 is the average.
  b 0.125 c 0.37

10 a $\frac{9}{25}$ or 0.36 b $\frac{111}{400}$ or 0.2775

11 a 0.84 b 0.9744

12 a $\frac{1}{4}$ b $\frac{3}{8}$

13 $\frac{27}{512}$

14 a 0.1 b 0.15 c 0.3

15 a i $\frac{k-5}{25}$ ii $\frac{2k-3}{25}$
   b $k = 8$; $\frac{49}{625}$

16 a i $\frac{1}{6}$ ii $\frac{2}{9}$
   b i 0 ii $\frac{1}{108}$

Exercise 4D

1 0.63

2 0.28

3 a 0.32 b 0.48

4 a i $\frac{0.35}{P(B)}$ ii $\frac{0.4}{P(C)}$
  b i 0.7 ii 0.5 iii 0.06

5 a

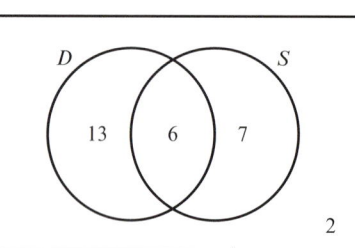

[Venn diagram: D and S intersecting, total = 28; D only = 13, intersection = 6, S only = 7, outside = 2]

	S	S'	Totals
D	6	13	19
D'	7	2	9
Totals	13	15	28

Answers

b No; $\frac{6}{28} \neq \frac{19}{28} \times \frac{13}{28}$

6 Yes; $\frac{20}{80} = \frac{32}{80} \times \frac{50}{80}$

7 a $P(A) = \frac{9}{16}, P(B) = \frac{3}{4}, P(A \cap B) = \frac{1}{2}$

b No; $\frac{1}{2} \neq \frac{9}{16} \times \frac{3}{4}$

c A and B both occur when, for example, 1 and 2 are rolled; $P(A \cap B) \neq 0$

8 a $P(X) = \frac{1}{4}, P(Y) = \frac{1}{3}, P(X \text{ and } Y) = \frac{1}{12}$

Yes; $\frac{1}{12} = \frac{1}{4} \times \frac{1}{3}$

b No; X and Y both occur when, for example, 1 and 5 are rolled; $P(X \cap Y) \neq 0$

9 $P(V) = \frac{1}{8}, P(W) = \frac{27}{64}, P(V \cap W) = \frac{1}{16}$

No; $\frac{1}{16} \neq \frac{1}{8} \times \frac{27}{64}$

10 a

	B	B'	Totals
M	60	48	108
M'	50	42	92
Totals	110	90	200

b Ownership is not independent of gender; e.g. for M and B: $\frac{60}{200} \neq \frac{108}{200} \times \frac{110}{200}$.

c Females 54.3%, males 55.6%. If ownership were independent of gender, these percentages would be equal.

11 $a = 1860, b = 4092, c = 1488$

12 Southbound vehicles; $\frac{36}{207} = \frac{54}{207} \times \frac{138}{207}$ or $\frac{18}{207} = \frac{54}{207} \times \frac{69}{207}$

Exercise 4E

1 a $\frac{2}{3}$ b $\frac{3}{4}$

2 a $\frac{3}{4}$ b $\frac{4}{7}$ c $\frac{10}{13}$

3 a $\frac{11}{19}$ b $\frac{12}{19}$

4 a i $\frac{5}{16}$ ii $\frac{12}{23}$

b Those who expressed an interest in exactly two (or more than one) career, or any other appropriate description.

5 a $\frac{20}{39}$ b $\frac{8}{39}$

6 a $\frac{1}{5}$ b $\frac{47}{57}$ or 0.825

7 a 10% of the staff are part-time females.

b $a = 0.2, b = 0.4, c = 0.3$

c i $\frac{4}{7}$ ii $\frac{3}{4}$ iii $\frac{4}{9}$

8 $\frac{3}{5}$ 9 $\frac{11}{21}$

10 a Proof

b $P(3) = 0.08, P(2) = 0.16, P(1) = 0.75$

c $\frac{25}{33}$ or 0.758

d $\frac{32}{107}$ or 0.299

Exercise 4F

1 a $\frac{3}{28}$ b $\frac{5}{14}$

2 $\frac{28}{55}$

3 a $\frac{7}{22}$ b $\frac{2}{33}$

4 a 0.027   b 229 or 230

5 a Two girls; $\frac{42}{132} > \frac{20}{132}$

b Equally likely; both $\frac{1}{66}$.

6 $\frac{1}{9}$

7 a $\frac{6}{19}$ or 0.316   b $\frac{1}{4}$

8 a $\frac{141}{400}$ or 0.3525   b $\frac{26}{47}$ or 0.553

9 a $\frac{1}{5}$ or 0.2   b $\frac{1}{3}$ or 0.333

c $\frac{7}{8}$ or 0.875

10 $\frac{18}{25}$ or 0.72

11 $\frac{9}{73}$ or 0.123

12 a $\frac{4}{7}$ or 0.571   b $\frac{13}{35}$ or 0.371

13 a $y = 0.44$ b $\frac{7}{20}$ or 0.35

14 a $x = 0.36$ b 0.812 or $\frac{272}{335}$

15 $\frac{23}{42}$ or 0.548

16 0.48

17 $\frac{16}{25}$ or 0.64

End-of-chapter review exercise 4

1. 53
2. $\frac{8}{11}$ or 0.727
3. a $\frac{104}{2185}$ or 0.0476

   b $\frac{1}{316}$ or 0.00316
4. $\frac{77}{248}$ or 0.310
5. $\frac{27}{40}$ or 0.675
6. a

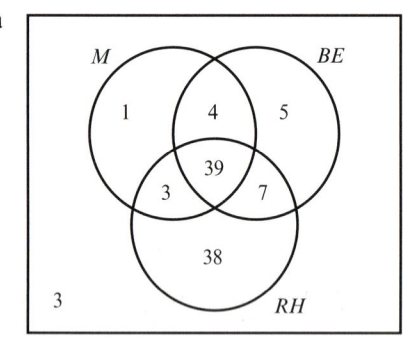

   (Venn diagram with three circles labelled M, BE, RH containing values: 1, 4, 5, 39, 3, 7, 38, and 3 outside)

   b i $\frac{14}{2475}$ or 0.00566

   ii $\frac{82}{495}$ or 0.166
7. a $0.3x + 0.7y = 0.034$ and $y = 2x$

   $x = 0.02, y = 0.04$

   b $\frac{16}{23}$ or 0.696
8. a $\frac{8}{9}$

   b $\frac{1}{4}$

   c $\frac{1}{3}$
9. i $\frac{8}{105}$ or 0.0762

   ii $\frac{2}{35}$ or 0.0571
10. i 0.85, 0.15 / 0.8, 0.2 / 0.4, 0.6 on branches with labels T, B / J, X / J, X.

    ii $\frac{17}{26}$ or 0.654
11. i $\frac{37}{85}$ or 0.435

    ii $\frac{19}{48}$ or 0.396

    iii Yes; P(high GDP and high birth rate) = 0

    iv $\frac{287}{666}$ or 0.431
12. a $\frac{4}{15}$

    b $\frac{9}{19}$
13. 0.198
14. a A and B both occur *or* it shows that $P(A \cap B) \neq 0$.

    b Only two of the 36 outcomes, (1, 3) and (3, 1), are favourable to A and to B.

    c No; $\frac{1}{18} \neq \frac{1}{6} \times \frac{1}{2}$ to show $P(A \cap B) \neq P(A) \times P(B)$.
15. 0.26
16. $x = 54$; 312 adults
17. a i $\frac{29}{34}$

    ii $\frac{3}{4}$

    b $\frac{13}{30}$
18. $\frac{17}{72}$
19. $\frac{8}{35}$
20. a i $\frac{1}{8}$

    ii $\frac{1}{4}$

    b 22

5 Permutations and combinations

Prerequisite knowledge

$P(A | B) = \frac{2}{5}$, $P(B | A) = \frac{2}{3}$

Exercise 5A

1. a 20 b 6 c 294

   d 162 e 224
2. a 10 b 9 c 4
3. a 11 b 15 c 22
4. E.g. $144 = \frac{9! \times 2!}{7!}$; $252 = \frac{7! \times 3!}{5!}$; $1\frac{1}{2} = \frac{15! \times 4!}{16!}$
5. $\frac{53!}{51!}$ cm^2

Answers

6 $\dfrac{25!}{22!} - \dfrac{8!}{5!}$ cm³

7 E.g. $\dfrac{9!}{5!(5!-4!+2!+2!)}$

Exercise 5B
1 720
2 a 8.07×10^{67}
  b 24
  c 6 227 020 800
3 a 2 b 720 c 40 320
4 a 24 b 6 c 5040
5 39 916 800
6 362 880
7 $n = 19$

Exercise 5C
1 a 120 b 360 c 45 360
  d 34 650 e 415 800
2 a 6 b 20
  c 60 d 15
3 a 6 b 1
  c 6435 d 99 768 240
4 First student is correct. Second student has treated them as two identical trees and three identical bushes.
5 a 1024
  b i 252 ii 386
6 One letter appears three times; another appears twice, and two other letters appear once each (e.g. pontoon, feeless, seekers, orderer).
7 a 10 b 50 c 1050

Exercise 5D
1 a 120
  b i 48 ii 72 iii 18
2 a 48 b 192 c 480
  d 144 e 0
3 2 : 1
4 a 80 640 b 241 920
5 a 3600 b 720 c 240
6 a 20 b 40
7 a 6 b 180 c 36

8 a 1 b 0
  c 8 d 20
9 $x > y + 1$ or $x \geqslant y + 2$ or equivalent

Exercise 5E
1 a 2520 b 3024
2 665 280
3 6840
4 a 182 b 196
5 a 60 b 240
6 a 272 b 132 c 140
7 a 60 480 b 1680
8 360
9 a 12 b 48
10 120 ways for $(r =) 3$ passengers to sit in $(n =) 6$ empty seats on a train, or use of $^5P_5$, $^5P_4$, or $^{120}P_1$.
11 a $r > \dfrac{1}{2}n$ b $k = \dfrac{n!}{r!(n-r)!}$
12 132 600
13 18 144
14 a 6 652 800
   b 3 024 000
   c 4 959 360

Exercise 5F
1 a 56 b 126
2 a 1960 b 980 c 121
3 a 2 598 960 b 845 000
4 a i 230 230 ii 230 230
  b $x = y + z$
5 16
6 161
7 a 120 b 34
  c 12 d 66
8 45
9 They can share the taxis in 56 ways, no matter which is occupied first.
10 a 184 756 b Two
   c 63 504 d 88 200
11 330
12 1 058 400
13 a 252 b 56 c 175

14 27 907 200
15 72
16 a 18   b 132

Exercise 5G

1 a $\frac{1}{3}$ b $\frac{2}{15}$ c $\frac{8}{15}$

2 a $\frac{21}{46}$ or 0.457

   b $\frac{27}{92}$ or 0.293

   c $\frac{3}{4}$

3 a 0.0260   b 0.197
4 0.0773
5 0.501
6 0.588
7 a $\frac{5}{16}$ b $\frac{1}{2}$

8 a $\frac{2}{3}$ b $\frac{1}{12}$ c $\frac{5}{12}$

9 a 0.331   b 0.937
10 a 50 400

   b i $\frac{1}{120}$ ii $\frac{1}{60}$

11 a $\frac{1}{84}$ or 0.0119   b $\frac{2}{9}$

12 0.290
13 a $a = 166, b = 274, c = 488$
   b 0.162
14 $\frac{28}{41}$ or 0.683
15 Six tags and three labels.
16 a $\frac{1}{3}, \frac{1}{2}, \frac{3}{5}$ and $\frac{2}{3}$ for $n = 2, 3, 4$ and 5.

   b $\frac{2}{n-1}$

End-of-chapter review exercise 5

1 a 30 240
   b 240
2 a 32 659 200
   b 8 467 200
3 $\frac{3}{28}$

4 a 1 000 000
   b i 0.01
      ii 0.0001
5 17 280
6 i   1 663 200
   ii 30 240
   iii 1 622 880
   iv 10
7 a 756 756
   b 72 072
8 a 330
   b 70
   c 265
9 91
10 87
11 a 10^9 or 1×10^9
   b $9^3 \times 10^6$ or 7.29×10^8
   c 9×10^8
   d $5^3 \times 10^6$ or 1.25×10^8
12 $\frac{9}{14}$
13 44 286
14 a 453 600
   b 86 400
15 $\frac{1}{10}$
16 a 11 values; 35
   b $\frac{11}{12}$
17 20; $\frac{2401}{2916}$ or 0.823
18 156
19 a i   648
      ii 104
   b 2700
20 i   50
   ii 18

Cross-topic review exercise 2

1 a 96
   b −71
   c $9\frac{1}{8}$ or 9.125

Answers

2   a   62

     b   Odd; $\dfrac{16}{31} > \dfrac{15}{31}$

3   30 856

4   a   i   48

         ii   24

     b   120

5   a   27

     b   i $\dfrac{1}{3}$

         ii $\dfrac{4}{9}$

6   1440

7   a   604 800

     b   8 467 200

8   a   134 596

     b   i $\dfrac{1}{24}$ ii $\dfrac{13}{24}$

     c   i $\dfrac{1}{19}$; more likely.

         ii $\dfrac{10}{19}$; less likely.

9   a   1287   b   45   c   270

10   a   3

     b   15

11   a   81

     b   15

12   a   i $\dfrac{14}{29}$ or 0.483

         ii $\dfrac{4}{29}$ or 0.138

     b   0.437

13   a

	P	P'	
S	30	10	40
S'	45	15	60
	75	25	100

      or appropriate Venn diagram.

     b   Yes; e.g. $\dfrac{30}{100} = \dfrac{40}{100} \times \dfrac{75}{100}$ to show $P(S \text{ and } P) = P(S) \times P(P)$.

14   a 3^{12} or 531 441

     b 3^{10} or 59 049

     c 3^{7} or 2187

15   a   3 628 800

     b   7 257 600

     c   39 916 800

     d   59 512 320

16   a   229 975 200

     b   0.75

17   a   i $\dfrac{8}{17}$ or 0.471

         ii $\dfrac{73}{153}$ or 0.477

         iii $\dfrac{32}{153}$ or 0.209

     b   The events 'being on the same side' and 'being in the same row' are not independent.

6 Probability distributions

Prerequisite knowledge

1   $P(D) = 0.11$

2   With replacement: $P(\text{both red}) = \dfrac{3}{6} \times \dfrac{3}{6} = \dfrac{1}{4}$ or 0.25

    Without replacement: $P(\text{both red}) = \dfrac{3}{6} \times \dfrac{2}{5} = \dfrac{1}{5}$ or 0.20

Exercise 6A

1

v	1	2	3
$P(V = v)$	0.4	0.4	0.2

2   $p = \dfrac{2}{13}; \dfrac{5}{13}$

3   a $50k^2 - 25k + 3 = 0;\ k = 0.2,\ k = 0.3$

     b $k = 0.3$ gives $P(W = 12) = -0.1$.

     c   0.14

4

s	0	1	2
$P(S = s)$	$\dfrac{4}{81}$	$\dfrac{28}{81}$	$\dfrac{49}{81}$

5   a   Proof

     b

r	0	1	2	3
$P(R = r)$	0.226	0.446	0.275	0.0527

     c   0.774

6   a   Proof

     b

v	0	1	2	3
$P(V = v)$	$\dfrac{24}{91}$	$\dfrac{45}{91}$	$\dfrac{20}{91}$	$\dfrac{2}{91}$

     c $\dfrac{69}{91}$

237

7   Number of red grapes selected (R); $R \in \{0, 1\}$
    Number of green grapes selected (G); $G \in \{4, 5\}$
    $R + G = 5$

8
d	0	1	2
$P(D = d)$	0.1	0.6	0.3

9
x	0	1	2
$P(X = x)$	0.4096	0.4608	0.1296

    Hair colour and handedness are independent.

10 **a** Proof

    b
x	2	3	4	5	6	7	8	10
$P(X = x)$	$\frac{1}{16}$	$\frac{2}{16}$	$\frac{3}{16}$	$\frac{2}{16}$	$\frac{3}{16}$	$\frac{2}{16}$	$\frac{2}{16}$	$\frac{1}{16}$

    $P(X > 6) = \frac{5}{16}$

11 **a** 0

    b
n	1	2	3	4
$P(N = n)$	$\frac{1}{14}$	$\frac{6}{14}$	$\frac{6}{14}$	$\frac{1}{14}$

    c Symmetrical

12 $k = \frac{1}{27}$

13 **a** $c = \frac{1}{86}$

    b $\frac{61}{86}$

14 **a** 0.374

    b $N = 0$ is more likely than $N = 4$;
    $P(N') > P(N)$ each time a book is selected.

15 **a** Proof

    b
x	0	1	2	3
$P(X = x)$	$\frac{1}{12}$	$\frac{4}{12}$	$\frac{4}{12}$	$\frac{3}{12}$

    $P(X \text{ is prime}) = \frac{7}{12}$

16 **a** P(heads) = 0.2

    b The number of tails obtained, but many others are possible, such as $2H$ and $0.5H$.
    $P(T > H) = 0.896$

17 **a**
s	1	2	3
$P(S = s)$	$\frac{17}{36}$	$\frac{9}{36}$	$\frac{10}{36}$

    b $\frac{1}{3}$

18 **a** $k = \frac{315}{1012}$ **b** $\frac{21}{46}$

Exercise 6B

1   $E(X) = 2.1$; $Var(X) = 0.93$

2   **a** $p = 0.2$
    b $E(Y) = 1.84$; $SD(Y) = 0.946$

3   $E(T) = 5$, $Var(T) = 11.5$

4   $m = 16$; $Var(V) = 31.3956$

5   $Var(R) = 831$

6   $a = 11$; $Var(W) = 79.8$

7   **a** $E(\text{grade}) = 3.54$; $SD(\text{grade}) = 1.20$; A smallish profit.
    $SD = 1.20$; variability of the profit.
    b $E(\text{grade}) = 2.46$, $SD(\text{grade}) = 1.20$
    Both are unchanged.

8   **a**
x	1	2	3	4	5	6	10	12	15	20	30
$P(X = x)$	$\frac{1}{36}$	$\frac{3}{36}$	$\frac{3}{36}$	$\frac{5}{36}$	$\frac{3}{36}$	$\frac{9}{36}$	$\frac{2}{36}$	$\frac{4}{36}$	$\frac{2}{36}$	$\frac{2}{36}$	$\frac{2}{36}$

    b $E(X) = 8\frac{5}{12}$; $P[X > E(X)] = \frac{1}{3}$
    c $Var(X) = 49\frac{41}{48}$ or 49.9

9   **a**
h	0	1	2	3
$P(H = h)$	0.343	0.441	0.189	0.027

    b 900 times

10 **a** $E(G) = 0.8$; $E(B) = 1.2$
    b $2 : 3$; It is the same as the ratio for the number of girls to boys in the class.
    c $Var(G) = 0.463$ or $\frac{336}{725}$

11 **a** Proof
    b $E(R) = 1.125$
    c $E(G) = 1.5$

12 **a** $340
    b If the successful repayment rate is below 70%.

13 **a** Proof   **b** $n = 35$

14 **a** 1, 2, 3, 5.

s	1	2	3	5
$P(S = s)$	$\frac{4}{12}$	$\frac{1}{12}$	$\frac{4}{12}$	$\frac{3}{12}$

    b $P(S > 2\frac{3}{4}) = \frac{7}{12}$

    c $Var(S) = 2\frac{17}{48}$

15 a Proof

b | x | 0 | 1 | 2 | 3 | 4 |
|---|---|---|---|---|---|
| $P(X=x)$ | $\frac{1}{256}$ | $\frac{12}{256}$ | $\frac{54}{256}$ | $\frac{108}{256}$ | $\frac{81}{256}$ |

$\frac{\text{Var}(X)}{\text{E}(X)} = \frac{1}{4}$

c The probability of not obtaining B with each spin.

End-of-chapter review exercise 6

1 $\left(k = \frac{9}{14}\right)$; $\text{E}(X) = 2\frac{5}{14}$ or 2.36
  $\text{Var}(X) = 1\frac{45}{196}$ or 1.23

2 a $q = 13$ or $q = 48$
  b 34

3 a $6675
  b 4.27

4 $\frac{6}{11}$

5 0.909

6 a i $\frac{2}{5}$
    ii $\frac{3}{5}$
  b | j | 0 | 1 | 2 |
|---|---|---|---|
| $P(J=j)$ | 0.3 | 0.6 | 0.1 |

7 a $S = 1$

b | s | 0 | 1 | 2 | 3 | 4 | 5 | 7 | 8 | 9 | 11 | 14 | 15 | 19 | 24 |
|---|---|---|---|---|---|---|---|---|---|---|---|---|---|---|
| $P(S=s)$ | $\frac{1}{36}$ | $\frac{13}{36}$ | $\frac{2}{36}$ | $\frac{3}{36}$ | $\frac{2}{36}$ | $\frac{2}{36}$ | $\frac{2}{36}$ | $\frac{1}{36}$ | $\frac{2}{36}$ | $\frac{2}{36}$ | $\frac{2}{36}$ | $\frac{1}{36}$ | $\frac{2}{36}$ | $\frac{1}{36}$ |

$\text{E}(S) = 5\frac{31}{36}$ or 5.86

8 a 0, 1, 2, 4, 5.

b | x | 0 | 1 | 2 | 4 | 5 |
|---|---|---|---|---|---|
| $P(X=x)$ | $\frac{1}{9}$ | $\frac{3}{9}$ | $\frac{2}{9}$ | $\frac{1}{9}$ | $\frac{2}{9}$ |

c $\frac{2}{3}$

d $a = \frac{\sqrt{7}}{2}$

9 a $b = 1$ or $b = 6$
  b $\frac{13}{30}$

10 i Proof

ii | Score | 0 | 2 | 4 | 6 |
|---|---|---|---|---|
| P(Score) | $\frac{24}{70}$ | $\frac{30}{70}$ | $\frac{13}{70}$ | $\frac{3}{70}$ |

iii $1\frac{6}{7}$ and 2.78

iv $\frac{2}{5}$ or 0.4

11 $\left(k = \frac{5}{3}\right)$; $P(Y > 4) = \frac{2}{3}$

12 $\frac{25}{28}$

13 a $x = 11$
   b $\frac{22}{127}$

14 i Proof

ii | x | 2 | 3 | 4 |
|---|---|---|---|
| $P(X=x)$ | $\frac{1}{6}$ | $\frac{1}{3}$ | $\frac{1}{2}$ |

iii $\frac{1}{3}$

15 i Proof

ii | x | 120 | 60 | 40 | 30 | 24 | 20 | $17\frac{1}{7}$ | 15 | $13\frac{1}{3}$ |
|---|---|---|---|---|---|---|---|---|---|
| $P(X=x)$ | $\frac{1}{45}$ | $\frac{2}{45}$ | $\frac{3}{45}$ | $\frac{4}{45}$ | $\frac{5}{45}$ | $\frac{6}{45}$ | $\frac{7}{45}$ | $\frac{8}{45}$ | $\frac{9}{45}$ |

iii $13\frac{1}{3}$ or 13.3

iv $\frac{4}{9}$ or 0.444

7 The binomial and geometric distributions

Prerequisite knowledge

1 105

2 $\frac{1}{64} + \frac{9}{64} + \frac{27}{64} + \frac{27}{64} = 1$

Exercise 7A

1 a 0.0016   b 0.4096
  c 0.0256   d 0.0272

2 a 0.0280   b 0.261
  c 0.710   d 0.552

3 a 0.0904   b 0.910
  c 0.163   d 0.969

4	a	0.121	b	0.000933	c	0.588
	d	0.403	e	0.499		
5	a	0.246	b	0.296		
6	0.0146					
7	0.254					
8	a	0.140	b	0.000684		
9	0.177					
10	a	0.599	b	0.257		
11	0.349					
12	a	0.291	b	0.648		
13	a	0.330	b	0.878		
14	a	0.15625 or $\frac{5}{32}$	b	0.578		
15	9		16	6		
17	16		18	23		
19	a	0.0098	b	$a = 208, b = 3$	c	68
20	a	$p = 0.5$; the probability of more than 5 m of rainfall in any given month of the monsoon season.
	b	The probability of more than 5 m of rainfall in any given month in the monsoon season is unlikely to be constant *or* Whether one month has more than 5 m of rainfall is unlikely to be independent of whether another has.
21	a	0.6561	b	0.227		
22	0.244					

Exercise 7B

1	a	1, 0.8 and 0.894	b	13.2, 5.94 and 2.44
	c	65.7, 53.874 and 7.34	d	14.1, 4.14 and 2.04
2	a	2 and 1.5	b	0.311	c	0.367
3	a	0.752	b	0.519
4	a	$n = 50, p = 0.4$	b	0.109
5	a	$n = 42, p = \frac{7}{12}$	b	0.0462
6	$n = 3, p = 0.9$

w	0	1	2	3
P(W = w)	0.001	0.027	0.243	0.729

7	a	E.g. X is not a discrete variable *or* there are more than two possible outcomes.
	b	E.g. Selections are not independent.
	c	E.g. X can only take the value 0 *or* X is not a variable.
8	$n = 18$; 0.364
9	$p = 0.75, k = 5157$

10	a	6.006
	b	5.93 and 5.93
	c	Proof
	d	0.197
11	a	46
	b	3.68
	c	i	0.566
		ii	0.320

Exercise 7C

1	a	0.0524	b	0.91808	c	0.4096
2	a	0.148	b	0.901	c	0.0672
3	a	0.125	b	0.875		
4	a	0.0465	b	0.482		
5	a	0.24	b	0.922	c	0.0280
6	a	i 0.032	ii	0.0016	b	0.2016
7	a	i 0.0315	ii	0.484	iii	0.440
	b	Faults occur independently and at random.
8	a	0.21	b	0.21	c	0.21
9	a	0.364	b	0.547		
10	0.0433
11	a	Not suitable; trials not identical (p not constant).
	b	Not suitable; success dependent on previous two letters typed *or* X cannot be equal to 1 or 2 or p is not constant.
	c	It is suitable.
	d	Not suitable; trials not identical (p not constant).
12	0.096
13	0.176
14	0.977 or $\frac{335}{343}$
15	a	0.0965 or $\frac{125}{1296}$	b	0.543
16	0.103

Exercise 7D

1	$2\frac{7}{9}$
2	5
3	$\frac{14}{81}$
4	Mode = 1, mean = 2
5	6 and 0.335
6	a	16	b	0.00366

7   a Thierry
    b $\frac{45}{784}$ or 0.0574
8   a With replacement, so that selections are independent.
    b i $\frac{27}{256}$ or 0.105
       ii $\frac{1}{16}$ or 0.0625
9   $E(X) = 500; b = 1001$
10 a Any representation of the following sequence.

	1st toss	2nd toss	3rd toss
Anouar	T	T	H
Zane	T	T	

    b $0.5^2 + 0.5^4 + 0.5^6 + 0.5^8 + ...$
    c $\frac{2}{3}$

End-of-chapter review exercise 7

1   $\left(\frac{n-1}{n}\right)^{n-1}$
2   a 0.147
    b 0.00678
3   i $\frac{1}{4}$
    ii 0.0791 or $\frac{81}{1024}$
    iii 0.09375 or $\frac{3}{32}$
4   a $\frac{4}{9}$
    b 0.394
5   a 0.59049
    b 0.40951
    c 0.242
6   a 10
    b 0.00772 or $\frac{5}{648}$
    c 0.00162
7   a 2^{-12}
    b 137×2^{-16}
8   a i   0.0706
       ii 0.0494
       iii 0.118
    b The students wear earphones independently and at random.
9   i 0.993
    ii $n = 22$

10 $k = 21$
11 $k = \left(\frac{3}{2}\right)^{n-2}$; $n = 10$
12 a 12
    b i   0.263
       ii 0.866
       iii 0.0199
13 i 0.735
    ii $n = 144; k = 6$
14 36 : 30 : 25

8 The normal distribution

Prerequisite knowledge
1   23.4 and 11.232
2   $n = 32, p = 0.35$

Exercise 8A
1   a False   b True   c False
    d False   e True   f False
2   a i $\sigma_P > \sigma_Q$
       ii Median for $P <$ median for Q.
       iii IQR for $P >$ IQR for Q.
    b i   Same as range of P.
       ii No; High values of W are more likely than low values or negatively skewed.
       iii
3

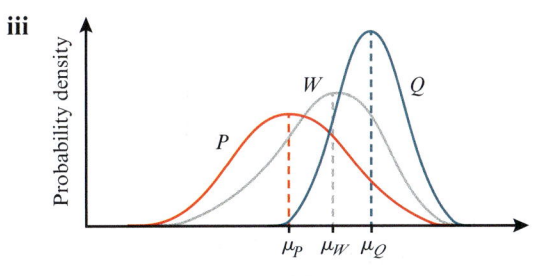

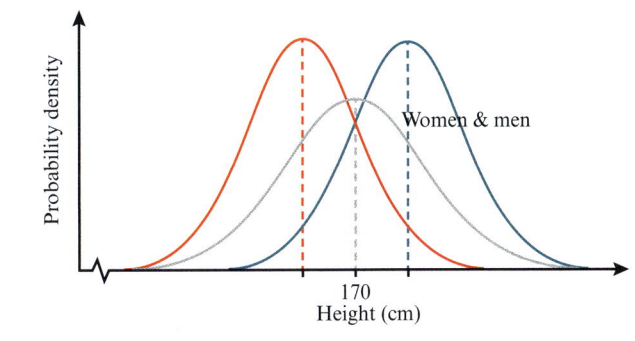

4 a

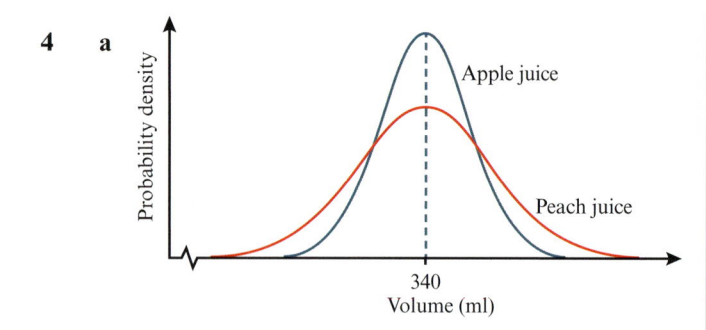

Peach juice curve wider and shorter than apple juice curve; equal areas; both symmetrical; both centred on 340 ml.

5 a

b USA curve wider, shorter and centred to the right of UK curve; equal areas; both symmetrical.

6 a Proof
  b $\sigma_x = 1.11 > \sigma_y = 0.663$

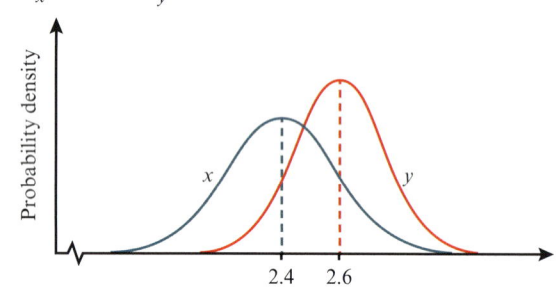

Exercise 8B

1 a 0.715   b 0.993
   c 0.937   d 0.531
   e 0.207   f 0.0224
   g 0.0401   h 0.495
   i 0.975   j 0.005

2 a 0.0606   b 0.380
   c 0.0400   d 0.0975
   e 0.190   f 0.211
   g 0.770   h 0.948
   i 0.719   j 0.066

3 a $k = 1.333$ b $k = 0.111$
   c $k = 0.600$ d $k = 1.884$
   e $k = -0.674$ f $k = -0.371$
   g $k = -1.473$ h $k = -0.380$
   i $k = 1.71$ j $k = 1.035$

4 a $c = 0.473$ b $c = 0.003$
   c $c = 2.10$ d $c = 1.245$
   e $c = -0.500$ f $c = -2.14$
   g $c = 3.09$ h $c = 1.96$
   i $c = 0.497$ j $c = -1.90$

Exercise 8C

1 a 0.726   b 0.191   c 0.629
2 a 0.919 and 0.0808   b 0.613 and 0.387
   c 0.964 and 0.0359   d 0.0467 and 0.953
   e 0.285 and 0.715   f 0.954
   g 0.423   h 0.319
   i 0.231   j 0.0994
3 a $a = 35.0$ b $b = 15.5$
   c $c = 18.5$ d $d = 23.6$
   e $e = 86.8$
4 a $f = 11.4$ b $g = 42.7$
   c $h = 9.80$ d $j = 17.5$
5 0.0513
6 0.933
7 $\sigma = 2.68$
8 $\mu = 12.6$
9 $\mu = 58.8, \sigma = 14.7$
10 $\mu = 93.8, \sigma = 63.8$
11 $\mu = 5, \sigma = 6.4; 0.0620$
12 $\mu = 7.08, \sigma = 1.95; 0.933$
13 $\mu = 5.78, \sigma = 2.13; 0.372$
14 0.831

Exercise 8D

1 0.662
2 a 0.191   b 74
3 a Small = 28.60%; medium = 49.95%; large = 21.45%
   b $k = 58.0$ or 58.1

4 $\mu = 7.57$
5 a 0.567   b 0.874   c 0.136
6 a $b = 240$ b 82.0 m
7 $9.09 \to 9$ days
8 5000
9 $\sigma = 3.33$
10 $\mu = 91.2$; 28.8%
11 $\sigma = 3.88$
12 a $\sigma = 1.83$ b 23
13 a $\mu = 25.0$ b $n = 1000$
14 a 0.683   b 0.0456
   c $\sigma = 1.64, \mu = 6.39$
15 a 0.950   b $n = 14$
16 a 0.659   b 0.189   c 0.257
17 a 0.284   b 0.0228   c 0.118

Exercise 8E

1 a Yes; $\mu = 12, \sigma^2 = 4.8$
   b No; $nq = 1.5 < 5$
   c Yes; $\mu = 5.2, \sigma^2 = 4.524$
   d No; $np = 3 < 5$
2 a $n = 209$ b $n = 34$
   c $n = 11$ d $n = 17$
3 B(56, 0.25)
4 0.837
5 0.844
6 a $p = 0.625$; Var(H) = 37.5
   b 0.0432
7 a Proof
   b 0.292; $np = 10 > 5$ and $nq = 30 > 5$
8 a 44   b 4.45   c 0.156
9 a i 0.187   ii 0.0118
   b E(X) = 1600; Var(X) = 320
   c 0.874
10 a i 0.105   ii 0.135
   b 0.145
11 0.0958
12 a 0.0729   b 0.877
13 a 0.239   b 0.0787
14 a 0.789   b 0.920
15 0.100
16 0.748
17 0.660

End-of-chapter review exercise 8

1 0.841
2 0.824
3 i 0.590
   ii $np = 24 > 5$ and $nq = 6 > 5$
4 0.287
5 0.239
6

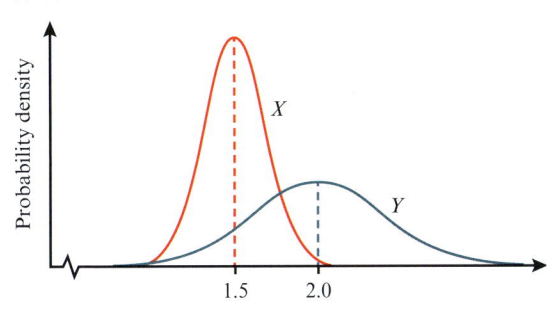

7 i 0.0350
   ii 0.471
   iii $k = 103$
8 a i 315 or 316
      ii 7350
      iii 0.840
   b 0.933
9 $\sigma = 2.35$
10 $\mu = 3.285$; 61.3%
11 5.69%
12 i 0.238
   ii $k = 116$
   iii 0.0910
13 a 0.408
   b 0.483
14 a $\mu = 17.5, \sigma^2 = 58.0$
   b i 17.0 min
      ii $38.2 \to 38$ days
15 a $\sigma = 7.24$
   b $k = 15.1$
16 0.936

Cross-topic review exercise 3

1 a i Proof
      ii E(X) = $\frac{3}{4}$, Var(X) = $\frac{63}{80}$ or 0.7875
      iii $\frac{399}{400}$ or 0.9975

b $\frac{13}{40}$ or 0.325

2   $\frac{19}{27}$

3   $\mu = 25.8, \sigma = 7.27$

4   0.0228

5   a $\sigma = 2.99$
    b   26.1% or 26.2%

6   a $\mu = 34.0$
    b   11.9%

7   a $\sigma = 2.70$
    b   0.276
    c   0.822 to 0.824

8   a
x	0	1	2	3
$P(X = x)$	$\frac{140}{285}$	$\frac{120}{285}$	$\frac{24}{285}$	$\frac{1}{285}$

    b   i $\frac{3}{5}$
        ii $\frac{24}{29}$ or 0.828

9   a $6\frac{2}{3}$
    b $\frac{2}{3}$

10 a $x = 5$
    b $\frac{209}{324}$ or 0.645

11 a   0.135
    b   0.253
    c $np = 9.35$ and $nq = 45.65$ are both greater than 5

12 a   Proof
    b   0.432

13 a $E(X) = 4$
    b   Proof
    c   0.315
    d $np = 8.4375 > 5$ and $nq = 11.5625 > 5$

14 a   i $\frac{1}{48} < p < \frac{47}{48}$
        ii $0 < p < \frac{1}{4}$ or $\frac{3}{4} < p < 1$
    b $\frac{1}{48} < p < \frac{1}{4}$ or $\frac{3}{4} < p < \frac{47}{48}$

Practice exam-style paper

1   a   1.72 m
    b   16.75

2   a $x = 0.154$; the value of $P(A \text{ and } B)$ or $P(A \cap B)$
    b $P(A \cap B) \neq 0$ or equivalent.
    c   Proof

3   a   36
    b $\frac{5}{9}$

4   a   0.35
    b $\frac{95}{137}$ or 0.693

5   a   11 km
    b   61
    c   10.8 km

6   a   12
    b   First trial
    c   0.176

7   a $x = 7, y = 4$
    b $b = 6.6$
    c   It is neither central nor representative or 8 of the 10 values are less than 89.

8   a
x	0	1	2
$P(X = x)$	$\frac{10}{28}$	$\frac{15}{28}$	$\frac{3}{28}$

    b $\frac{45}{112}$ or 0.402
    c $\frac{11}{15}$

Glossary

The following abbreviation and symbols are used in this book.

	Meaning
No.	Number of
≈	is approximately equal to
≠	is not equal to
∝	is proportional to
∴	therefore
≡	is identical to

A

Arrangements: see permutations

Average: any of the measures of central tendency, including the mean, median and mode

B

Binomial distribution: a discrete probability distribution of the possible number of successful outcomes in a finite number of independent trials, where the probability of success in each trial is the same

C

Categorical data: see qualitative data

Class: a set of values between a lower boundary and an upper boundary

Class boundaries: the two values (lower and upper) between which all the values in a class of data lie

Class interval: the range of values from the lower boundary to the upper boundary of a class

Class mid-value (or midpoint): the value exactly half-way between the lower boundary and the upper boundary of a class

Class width: the difference between the upper boundary and the lower boundary of a class

Coded: adjusted throughout by the same amount and/or by the same factor

Combinations: the different selections that can be made from a set of objects

Complement: a number or quantity of something required to make a complete set

Continuity correction: an adjustment made when a discrete distribution is approximated by a continuous distribution

Continuous data: data that can take any value, possibly within a limited range

Cumulative frequency: the total frequency of all values less than a particular value

Cumulative frequency graph: a graphical representation of the number of readings below a given value made by plotting cumulative frequencies against upper class boundaries for all intervals

D

Dependent (events): events that cannot occur without being affected by the occurrence of each other

Discrete data: data that can take only certain values

E

Elementary event: an outcome of an experiment

Equiprobable: events or outcomes that are equally likely to occur

Expectation: the expected number of times an event occurs

Extreme value: an observation that lies an abnormal distance from other values in a set of data

F

Factorial: the product of all positive integers less than or equal to any chosen positive integer

Fair: not favouring any particular outcome, object or person

Favourable: leading to the occurrence of a required event

Frequency: the number of times a particular value occurs

Frequency density: frequency per standard interval

G

Geometric distribution: a discrete probability distribution of the possible number of trials required to obtain the first successful outcome in an infinite number of independent trials, where the probability of success in each trial is the same

Grouped frequency table: a frequency table in which values are grouped into classes

H

Histogram: a diagram consisting of touching columns whose areas are proportional to frequencies

I

Independent (events): events that can occur without being affected by the occurrence of each other

Interquartile range: the range of the middle half of the values in a set of data; the numerical difference between the upper quartile and the lower quartile

K

Key: a note that explains the meaning of each value in a diagram

L

Lower and upper boundary: the smallest and largest values that can exist in a class of continuous data

M

Mathematical model: a description of a system using mathematical concepts and language

Mean: the sum of a set of values divided by the number of values

Median: the number in the middle of an ordered set of values

Modal class: the class of values with the highest frequency density

Mode: the value that occurs most frequently

Mutually exclusive (events): events that cannot occur at the same time because they have no common favourable outcomes

N

Normal curve: a symmetrical, bell-shaped curve

Normal distribution: a function that represents the probability distribution of particular continuous random variables as a symmetrical bell-shaped graph

O

Ordered data: data arranged from smallest to largest (ascending) or largest to smallest (descending)

Outliers: extreme values; observations that lie an abnormal distance from other values in a set of data

P

Parameters: the fixed values that define the distribution of a variable

PDF: see probability density function

Permutations: the different orders in which objects can be selected and placed

Probabilities: measurements on a scale of 0 to 1 of the likelihood that an event occurs

Probability density function (PDF): a graph illustrating the probabilities for values of a continuous random variable

Probability distribution: a display of all the possible values of a variable and their corresponding probabilities

Q

Qualitative data: data that take non-numerical values

Quantitative data: data that take numerical values

Quartile: any of three measures that divide a set of data into four equal parts

R

Random: occurring by chance and without bias

Range: the numerical difference between the largest and smallest values in a set of data

Raw data: numerical facts and other pieces of information in their original form

Relative frequency: the proportion of trials in which a particular event occurs

S

Selection: an item or number of items that are chosen

Skewed: unsymmetrical

Standard deviation: a measure of spread based on how far the data values are from the mean; the square root of the variance

Standard normal variable: the normally distributed variable, Z, with mean 0 and variance 1

Stem-and-leaf diagram: a type of table for displaying ordered discrete data in rows with intervals of equal widths

Summarise: to give an accurate general description

T

Trial: one of a number of repeated experiments

U

Unbiased: not favouring any particular outcome, object or person

V

Variance: the mean squared deviation from the mean; the square of the standard deviation

Variation: dispersion; a measure of how widely spread out a set of data values is

Index

$\Phi(z)$ 195–9
   table of values 222

addition law, mutually exclusive events 94–5
averages *see* measures of central tendency

bar charts 13
binomial distribution 166
   expectation 173–4
   normal approximation 208–12
   variance 173–4
binomial expansions 167–8
box-and-whisker diagrams (box plots) 60

categorical (qualitative) data 2
   data representation 20
central tendency, measures of *see* measures of central tendency
class boundaries 6, 7
class frequency 7, 8
class widths 6, 7
classes
   histograms 6
   stem-and-leaf diagrams 3, 4
coded data 37–8, 40–1
   standardising a normal distribution 200–3
   variance and standard deviation 75–80
combinations 123, 135
   ^nCr notation 135–6
   problem solving 138–40
combined datasets
   mean 31–2, 73
   variance and standard deviation 72–3
complement of an event 92
conditional probability 108–9
   and dependent events 112–15
   and independence 111–12
continuity corrections 209–10
continuous data 3
   cumulative frequency graphs 13–15
   histograms 6–9
continuous random variables 188–9
   normal distribution 190–206
   probability density functions 189–90

cumulative frequency 13
cumulative frequency graphs 13–15
   estimation of the median 43–4
   interquartile range 58–9

data representation
   box-and-whisker diagrams 60
   comparing different methods 20
   cumulative frequency graphs 13–15
   histograms 6–9
   stem-and-leaf diagrams 3–4
data types 2–3
de Moivre, Abraham 208
dependent events 112–15
deviation 65
   see also standard deviation
discrete data 2–3
   stem-and-leaf diagrams 3–4
discrete random variables 150
   binomial distribution 166–74
   expectation 156–7, 158
   geometric distribution 166, 175–82
   probability distributions 150–2
   variance 157–8

elementary events (outcomes) 91
equiprobable events 91
errors 188
events 91
   dependent 112–15
   exhaustive 92
   independent 100–2, 111–12
   mutually exclusive 94–5
expectation 92–3
   of the binomial distribution 173–4
   of a discrete random variable 156–7, 158
   of the geometric distribution 180–1
   see also mean

factorial function 124–5
fair (unbiased) selection 91
Fermat's Last Theorem 156
frequency density 7–9

Gauss, Carl Friedrich 188, 205, 208
geometric distribution 166, 175–8
   expectation 180–1
   mode 180

grouped data
   mean 30–1
   variance and standard deviation 66, 67–8
grouped frequency tables 6
   estimation of the mean 32–3

height variation 64
histograms 6–9
   modal class 28
   use in image processing 13

independent events 100–2
   application of the multiplication law 105–6
   and conditional probability 111–12
interquartile range 56
   box-and-whisker diagrams 60
   comparison with standard deviation 69
   grouped data 58–9
   ungrouped data 56–7

mathematical models 166
   using the normal distribution 205–6
mean 27, 30–1, 44
   of the binomial distribution 173–4
   of coded data 37–8, 40–1
   of combined datasets 31–2, 73
   of a discrete random variable 156–7, 158
   of the geometric distribution 180–1
   from grouped frequency tables 32–3
   of a normal distribution 190–1, 193
measures of central tendency 27
   choosing an appropriate average 44–6
   effect of extreme values 45
   historical background 44
   mean 30–41
   median 42–4
   mode and modal class 28–9
   for skewed data 45–6
   see also mean; median; mode
measures of variation 55
   coded data 75–80
   interquartile range and percentiles 56–9
   range 55–6

measures of variation (*Cont.*)
   variance and standard deviation 65–9, 72–3
   see also standard deviation; variance
median 27, 42–3, 44, 56
   estimation from a cumulative frequency graph 43–4
modal class 28–9
mode 27, 44
   of the geometric distribution 180
multiplication law for independent events 100–2
   application of 105–6
multiplication law of probability 112–15
mutually exclusive events 94–5

normal curve 190–1
normal distribution 188, 193
   approximation to the binomial distribution 208–12
   modelling with 205–6
   properties of 194
   standard normal variable (Z) 195–9
   standardising 200–3
   tables of values 222

parameters
   of a binomial distribution 167
   of a geometric distribution 176
   of a normal distribution 193
Pascal's triangle 168, 172
percentiles 58–9
permutations 123, 134
   of *n* distinct objects 125–6
   of *n* distinct objects with restrictions 129–31
   of *n* objects with repetitions 127–8
   ^nP_n notation 125
   ^nP_r notation 132
   problem solving 138–40
   of *r* objects from *n* objects 132–3

possibility diagrams (outcome spaces) 101
possibility space 95
probability 91
   addition law 94–5
   conditional 108–9, 111–15
   dependent events 112–15
   experiments, events and outcomes 91–3
   independent events 100–2, 111–12
   multiplication law for independent events 100–2, 105–6
   multiplication law of probability 112–15
   mutually exclusive events 94–5
   Venn diagrams 95–7
probability density functions (PDFs) 189–90
probability distributions 150–2
   binomial distribution 166–74
   geometric distribution 175–82
   normal distribution 190–206

qualitative (categorical) data 2
   data representation 20
quantitative data 2–3
quartiles 56
   grouped data 58–9
   ungrouped data 56–7

random selection 91–2
range 55–6
repetitions, permutations with 127–8
restrictions, permutations with 129–31

selection, random 91–2
set notation 95
sigma (Σ) notation 30
skewed data
   box-and-whisker diagrams 60
   measures of central tendency 45–6

skewed distributions 190
standard deviation 65–8
   of the binomial distribution 173–4
   calculation from totals 72
   coded data 75–80
   of combined datasets 72–3
   comparison with interquartile range 69
   of a discrete random variable 158
   of a normal distribution 190–1, 194
standard normal variable (Z) 195–9
standardising a normal distribution 200–3
stem-and-leaf diagrams 3–4
   interquartile range 57
   median 42

tree diagrams
   for independent events 100–1
   for permutations 125–6
trials 92–3

variables
   notation 150
   see also continuous random variables; discrete random variables
variance 65–8
   of the binomial distribution 173–4
   calculation from totals 72
   of coded data 75–80
   of combined datasets 72–3
   of a discrete random variable 157–8
   equivalence of two formulae for 81
   of a normal distribution 193
variation 55
see also measures of variation
Venn diagrams 95–7

Wiles, Andrew 156

Z (standard normal variable) 195–9